Charles Davies

The metric System

Considered with Reference to its Introduction into the United States

Charles Davies

The metric System
Considered with Reference to its Introduction into the United States

ISBN/EAN: 9783337187095

Printed in Europe, USA, Canada, Australia, Japan

Cover: Foto ©berggeist007 / pixelio.de

More available books at **www.hansebooks.com**

THE
METRIC SYSTEM,

CONSIDERED WITH REFERENCE TO ITS INTRODUCTION INTO
THE UNITED STATES;

EMBRACING

THE REPORTS OF THE HON. JOHN QUINCY ADAMS, AND THE LECTURE OF SIR JOHN HERSCHEL.

BY CHARLES DAVIES, LL. D.,

Chairman of the Committee on Coins, Weights and Measures of the University Convocation of the State of New York.

A. S. BARNES AND COMPANY,
NEW YORK AND CHICAGO.
1871.

DAVIES'
COURSE OF MATHEMATICS.

Davies' Primary Arithmetic and Table-Book—Designed for Beginners; containing the elementary tables of Addition, Subtraction, Multiplication, Division, and **Denominate** Numbers; with a large number of easy and practical questions, both mental and written.

Davies' First Lessons in Arithmetic—Combining the Oral Method with the Method of Teaching the Combinations of Figures by Sight.

Davies' Intellectual Arithmetic—An Analysis of the Science of Numbers, with especial reference to Mental Training and Development.

Davies' New School Arithmetic—Analytical and Practical.

Key to Davies' New School Arithmetic.

Davies' Grammar of Arithmetic—An Analysis of the Language of Numbers and the Science of Figures.

Davies' University Arithmetic—Embracing the Science of Numbers, and their Applications according to the most Improved Methods of Analysis and Cancellation.

Key to Davies' University Arithmetic.

Davies' Elementary Algebra—Embracing the First Principles of the Science.

Key to Davies' Elementary Algebra.

Davies' Elementary Geometry and Trigonometry—With Applications in Mensuration.

Davies' Practical Mathematics—With Drawing and Mensuration applied to the Mechanic Arts.

Davies' University Algebra—Embracing a Logical Development of the Science, with graded examples.

Davies' Bourdon's Algebra—Including Sturm's and Horner's Theorems, and practical examples.

Davies' Legendre's Geometry and Trigonometry—Revised and adapted to the course of Mathematical Instruction in the United States.

Davies' Elements of Surveying and Levelling—Containing descriptions of the Instruments and necessary Tables.

Davies' Analytical Geometry—Embracing the Equations of the Point, the Straight Line, the Conic Sections, and Surfaces of the first and second order.

Davies' Differential and Integral Calculus.

Davies' Descriptive Geometry—With its application to Spherical Trigonometry, Spherical Projections, and Warped Surfaces.

Davies' Shades, Shadows, and Perspective.

Davies' Logic and Utility of Mathematics—With the best Methods of Instruction Explained and Illustrated.

Davies' and Peck's Mathematical Dictionary and Cyclopedia of Mathematical Science—Comprising Definitions of all the terms employed in Mathematics—an Analysis of each Branch, and of the whole, as forming a single Science.

PREFACE.

It is, perhaps, not generally known that the Metric System of Weights and Measures has been adopted, permissively, by the Congress of the United States, and that any Merchant, Mechanic, or Tradesman may, if he pleases, in strict conformity to law, render all his bills and keep his accounts according to that system.

The mind of the civilized world, brought into sympathy and close connection by the Wires of the Telegraph, is now earnestly directed to the question of uniformity in the language of business relations; and the Metric System of France is presented as a means of effecting such uniformity.

In May, 1866, the following bill was passed by the Congress of the United States:

A Bill to Authorize the Use of the Metric System of Weights and Measures.

Be it enacted by the Senate and House of Representatives of the United States of America in Congress assembled, That from and after the passage of this act, it shall be lawful throughout the United States of America to employ the weights and measures of the Metric System; and no contract, or dealing, or pleading in any court, shall be deemed invalid or liable to objection, because the weights or measures expressed or referred to therein are weights or measures of the Metric System.

SEC. 2. *And be it further enacted,* That the tables in the schedule hereto annexed shall be recognized in the construction of contracts, and in all legal proceedings, as establishing, in terms of the weights and measures now in use in the United States, the equivalents of the weights and measures expressed therein in terms of the Metric System; and said tables may be lawfully used for computing, determining, and expressing in customary weights and measures the weights and measures of the Metric System.

———

At the meeting of the University Convocation of the State of New York, at Albany, in the summer of 1866, the Hon. John A. Kasson, Member of Congress and Chairman of the Committee of the House of Representatives, on a "Uniform System of Coinage, Weights, and Measures," called the attention of the members to the action of Congress, and earnestly asked such attention to the subject as in their judgment might seem wise and proper.

A Committee, consisting of the Chancellor, **J. V. L. Pruyn,** Professor Charles Davies, and Regent Robert S.

Hale, was appointed, and instructed to report "What measures, if any, the Convocation should adopt in regard to a Uniform System of Weights and Measures."

The duty of collecting the materials for the report of the Committee was assigned, by the Chairman, to the second member named, and it seemed to be the unanimous opinion of the Committee that a report would be made, favorable to the introduction of the system into general use. On examination, however, it did not appear to the Committee, and especially to the one who had been charged with making the examinations, that the Convocation should commit itself, hastily, to the great and radical changes which the introduction of the Metric System would occasion. At the meeting of the Convocation in 1869, the Committee made a partial report, and explained, very fully, the changes which an examination of the subject had produced on the minds of some of its members, whereupon the Committee was discharged and a new Committee appointed, composed of Professor Charles Davies, Robert S. Hale, and Professor James B. Thompson. Professor Thompson has not acted with the Committee, and is, of course, not responsible for its doings.

The Committee have given the subject a very full and careful examination. They have shared the enthusiasm which the hope of a common Currency, a common unit of Weight, and a common unit of Measure, for all nations, has awakened throughout the world. They honor the French nation for having taken the first step in so great an undertaking. But in their judgment, the adoption of the Metric System, without modifications, and the entire obliteration of every unit of weight and measure which now form the warp of our language and the base of our traffic and commerce, while it is yet uncertain how far it may be adopted by other nations, would be most unwise.

We must not forget that the introduction of the Metric System carries with it the necessity of abandoning our own Saxon, and introducing a language entirely foreign, and which the masses will be obliged to use.

The report of the International Committee of the Paris Universal Exposition, states: "It will be observed that the French orthography is retained throughout. This is done for uniformity, and to avoid the word *gram*. * * * * The orthography should be retained for the sake of uniformity also. The Metric System is destined to be generally international. The names and orthography of all its divisions should be equally so. For this reason, alone, we should refrain from Anglicizing the French names."

The Committee have been most anxious to state all the facts of the case, in their regular order; and hence, they have given in Part I. the Metric System itself, with its weights and measures, which are its proper supplements.

In Part II. they have sought to give a fair and just analysis of the Metric System, considered specially with reference to its connection with our systems of public instruction.

Part III. is the able and extraordinary report of Mr.

John Quincy Adams. He examined the whole subject with the minuteness and accuracy of mathematical science —with the keen sagacity of statesmanship, and the profound wisdom of philosophy. To that report nothing can be added, and from it nothing should be taken away. Hence, the Committee have published it in full, that the public, and especially the teachers of the country, may understand the entire subject, in all its phases and in all its relations.

Part IV.. is the lecture of Sir John Herschel, on the Pendulum, the Yard, and the Metre, regarded as a standard. In this lecture the subject is examined from the standpoint of exact science, and in this regard it is well worthy of attention.

The Committee, in compliance with a vote of the Convocation, submit to the public not only the results of their own labors, but also the more important essays forming Parts III. and IV., in the hope that all may contribute to the advancement of science, the diffusion of knowledge, and to the final adoption, by all nations, of a common system of Weights and Measures

CONTENTS.

PART I.

METRIC SYSTEM.

THE primary base, in this system, for all denominations of weights and measures, is the one-ten-millionth part of the distance from the equator to the pole, measured on the earth's surface. It is called a METRE, and is equal to 39.37 inches, very nearly.

The change from the base, in all the denominations, is according to the decimal scale of tens: that is, the units increase ten times, at each step, in the ascending scale, and decrease ten times, at each step, in the descending scale.

MEASURES OF LENGTH.

Base, 1 metre = 39.37 inches, nearly.

Table.

Ascending Scale.					Descending Scale.		
1 Myriametre.	1 Kilometre.	1 Hectometre.	1 Decametre.	1 METRE. UNIT.	1 Decimetre.	1 Centimetre.	1 Millimetre.

1*

The names, in the ascending scale, are formed by prefixing to the base, Metre, the words, Deca (ten), Hecto (one hundred), Kilo (one thousand), Myria (ten thousand), from the Greek numerals; and in the descending scale, by prefixing Deci (tenth), Centi (hundredth), Milli (thousandth), from the Latin numerals. Hence, the name of a unit indicates whether it is greater or less than the standard; and, also, how many times. The table is thus read:

10 millimetres	make	1 centimetre.
10 centimetres	make	1 decimetre.
10 decimetres	make	1 METRE.
10 METRES	make	1 decametre.
10 decametres	make	1 hectometre.
10 hectometres	make	1 kilometre.
10 kilometres	make	1 myriametre.

Table of Equivalents.

Myriametre.	Kilometre.	Hectometre.	Decametre.	METRE.	Decimetre.	Centimetre.	Millimetre.
.	.	.	.	.	.	1 =	10
.	.	.	.	. =	1 =	10 =	100
.	.	.	.	1 =	10 =	100 =	1,000
.	.	. =	1 =	10 =	100 =	1,000 =	10,000
.	.	1 =	10 =	100 =	1,000 =	10,000 =	100,000
.	1 =	10 =	100 =	1,000 =	10,000 =	100,000 =	1,000,000
1 =	10 =	100 =	1,000 =	10,000 =	100,000 =	1,000,000 =	10,000,000

Table of Equivalents in English Measure.

1 Millimetre	= 0.0394 inches, nearly.
1 Centimetre	= 0.3937 "
1 Decimetre	= 3.9370 "

<pre>
1 METRE   = 39.37 in. = 3.280833 ft.
1 Decametre   = 32 ft. 9.7 in.
1 Hectometre = 19 rd. 14 ft. 7 in.
1 Kilometre   = 4 fur. 38 rd. 13 ft. 3 in.
1 Myriametre = 6 mi. 1 fur. 28 rd. 6 ft. 4 in.
</pre>

Besides a clear apprehension of the length of the base, 1 metre, it is well to consider the length of the largest unit, the myriametre, equal to nearly 6 and one-fourth miles; and also the length of the smallest unit, the millimetre, about four-hundredths of an inch. Compare also, each of the smaller measures, the decimetre and centimetre, with the inch.

When, in the metric system, the value of any single unit is fixed in the mind, the values of all the others may be readily apprehended, since they always arise from multiplying or dividing by 10.

NOTE.—In all the tables, the UNIT is in small capitals, and should be constantly referred to.

Methods of Reading.

The number 25365.897 metres, is read, in English,

Twenty-five thousand three hundred and sixty-five metres, and 897 thousandths of a metre. But in the language of the metric system, it may be read,

Two myriametres, 5 kilometres, 3 hectometres, 6 decametres, 5 metres, 8 decimetres, 9 centimetres, and 7 millimetres. It may also be read, beginning with the lowest denomination, 7 millimetres, 9 centimetres, etc., etc.

In reading, remember that the unit of any place is *one-tenth* of the unit in the place next at the left, and ten times as great as the unit of the place next at the right. Hence, the change from one unit to another, and the methods of reduction and reading, are identical with those in the system of decimal currency.

1. Write, numerate, and read, five hundred and ninety-six hectometres.

2. Write, numerate, and read, eighty-nine thousand and forty-one centimetres.

MEASURES OF SURFACES, OR SQUARE MEASURE.

Base, 1 Are = the square whose side is 10 metres.
= 119.6 square yards, nearly.
= 4 perches or square rods, nearly.

The unit of surface is a square whose side is 10 metres. It is called an ARE, and is equal to 100 square metres.

1 Hectare.   1 ARE. UNIT.   1 Centiare.

The table is thus read:

100 centiares	make	1 ARE.
100 ares	make	1 hectare

Table of Equivalents.

Hectare.	Are.	Centiare
	1 =	100
1 =	100 =	10,000

Equivalents in acres, roods, and perches.

> 1 Centiare = 1.195985 sq. yards, nearly.
> 1 ARE   = 3.95367 perches.
> 1 Hectare = 2A. 1R. 35.367P.

MEASURES OF VOLUMES.

Base, 1 Litre = the cube on the decimetre.
= 61.023378 cubic inches.
= a little more than a wine quart.

The unit for the measure of volume is the cube whose edge is one-tenth of the metre—that is, a cube whose edge is 3.937 inches. This cube is called a LITRE, and is one-thousandth part of the cube constructed on the metre, as an edge.

Ascending Scale.			Descending Scale.			
1 Kilolitre, or Stere.	1 Hectolitre.	1 Decalitre.	1 LITRE. UNIT.	1 Decilitre.	1 Centilitre.	1 Millilitre.

The table is thus read:

10 millilitres	make	1 centilitre.
10 centilitres	make	1 decilitre.
10 decilitres	make	1 litre.
10 litres	make	1 decalitre.
10 decalitres	make	1 hectolitre.
10 hectolitres	make	1 kilolitre, or stere.

Table of Equivalents.

Kilolitre or Stere.	Hectolitre.	Decalitre.	Litre.	Decilitre.	Centilitre.	Millilitre.
.	.	.	.	.	1 =	10
.	.	.	.	1 =	10 =	100
.	.	.	1 =	10 =	100 =	1,000
.	.	1 =	10 =	100 =	1,000 =	10,000
.	1 =	10 =	100 =	1,000 =	10,000 =	100,000
1 =	10 =	100 =	1,000 =	10,000 =	100,000 =	1,000,000

NOTE.—The kilolitre, or stere, is the cube constructed on the metre, as an edge. Hence, the litre is one-thousandth part of the kilolitre.

Equivalents in Cubic Measure.

1 millilitre	=	.061023 cubic inches.
1 centilitre	=	.610234 cubic inches.
1 decilitre	=	6.102338 cubic inches.
1 LITRE	=	61.023378 cubic inches.
1 decalitre	=	610.233779 cubic inches.
1 hectolitre	=	6102.337795 cu. in. = 3.5314454 cu. ft.
1 kilolitre, or stere	=	61023.377953 cu. in. = 35.314454 cu. ft.

NOTE.—Law of change in the units, and methods of reading, are the same as in Linear Measure.

DRY MEASURE.

EQUIVALENTS IN THE WINCHESTER BUSHEL.

Since 1 bushel = 2150.4 cu. in. ; 1 pk. = 537.6 cu. in.; 1 qt. = 67.2 cu. in. 1 pt. = 33.6 cu. in.; therefore,

1 millilitre	= .001816 pints.
1 centilitre	= .018160 pints.
1 decilitre	= .181606 pints.
1 LITRE	= 1.816060 pints.
1 decalitre	= 1 pk. 1.8083000 qt.
1 hectolitre	= 2 bu. 3 pk. 2 qt. 1.6 pt.
1 kilolitre, or stere	= 28 bu. 1 pk. 4 qt.

NOTE.—The litre, or standard, is a little less than 1 quart, and the stere, nearly 30 Winchester bushels.

LIQUID MEASURE.

EQUIVALENTS IN THE WINE GALLON.

Since 1 wine gallon contains 231 cubic inches, 1 quart will contain 37.75 cubic inches; 1 pint, 28.874 cubic inches; and 1 gill, 7.21875 cubic inches; we have,

1 millilitre	= 0.008453 gils.
1 centilitre	= 8.084534 gils.
1 decilitre	= 0.845320 gils.
1 LITRE	= 1 qt. 1.1335 pt.
1 decalitre	= 2 gal. 2 qt. 1.1335 pt.
1 hectolitre	= 26 gal. 1 qt. 1 pt. 1.34 gils.
1 kilolitre, or stere	= 1 tun, 12 gal. 1 pt. 1.44 gils.

WEIGHTS.

Base, 1 Gramme = weight of a cubic centimetre of rain-water.
= 15.432 grains, Troy, nearly.
= .0352746 ounces, Avoirdupois, nearly.

The unit of weight is also equal to the one-millionth part of the weight of a cubic metre of pure rain-water, weighed in vacuum. It is called a GRAMME, and is equal to 15.432 grains, Troy, which is equal to .0352746 ounces, Avoirdupois, very nearly.

<table>
<tr><td></td><td colspan="6" align="center">Ascending Scale.</td><td></td><td colspan="3" align="center">Descending Scale.</td></tr>
<tr>
<td>1 Millier, tonneau.</td>
<td>1 Quintal.</td>
<td>1 Myriagramme.</td>
<td>1 Kilogramme.</td>
<td>1 Hectogramme.</td>
<td>1 Decagramme.</td>
<td>1 GRAMME. UNIT.</td>
<td>1 Decigramme.</td>
<td>1 Centigramme.</td>
<td>1 Milligramme.</td>
</tr>
</table>

The table is thus read:

10 milligrammes	make	1 centigramme.
10 centigrammes	make	1 decigramme.
10 decigrammes	make	1 GRAMME.
10 GRAMMES	make	1 decagramme.
10 decagrammes	make	1 hectogramme.
10 hectogrammes	make	1 kilogramme.
10 kilogrammes	make	1 myriagramme.
10 myriagrammes	make	1 quintal.
10 quintals	make	1 millier, or tonneau.

Table of Equivalents.

Millier, or tonneau.	Quintal.	Myriagramme.	Kilogramme.	Hectogramme.	Decagramme.	GRAMME.	Decigramme.	Centigramme.	Milligramme.
								1=	10
							1=	10=	100
						1=	10=	100=	1,000
					1=	10=	100=	1,000=	10,000
				1=	10=	100=	1,000=	10,000=	100,000
			1=	10=	100=	1,000=	10,000=	100,000=	1,000,000
		1=	10=	100=	1,000=	10,000=	100,009=	1,000,000=	10,000,000
	1=	10=	100=	1,000=	10,000=	100,000=	1,000,000=	10,000,000=	100,000,000
1=	10=	100=	1,000=	10,000=	100,000=	1,000,000=	10,000,000=	100,000,000=	1,000,000,000

Equivalents in Avoirdupois and Troy Weights.

1 milligramme	=	0.0154	grains, Troy.
1 centigramme	=	0.1543	grains, "
1 decagramme	=	1.5432	grains, "
1 GRAMME	=	15.4327	grains, "
1 decagramme	=	0.3527	ounces, Avoirdupois.
1 hectogramme	=	3.5274	ounces, "
1 kilogramme	=	2.2046	pounds, "
1 myriagramme	=	22.046	pounds, "
1 quintal	=	220.46	pounds, "
1 millier, or tonneau	=	2204.6	pounds, "

NOTE.—Law of change in the units, and methods of reading, the same as in Linear Measure.

NATURE OF THE METRIC SYSTEM.

The Metric system is based on the METRE. From the metre, three other units are derived; and the four constitute the primary units of the system. They are:

METRE = 39.37 inches, nearly: unit of length.

ARE = a square on 10 metres: unit of surface.

LITRE = a cube whose edge is a decimetre: unit of volume.

GRAMME = the weight of a cube of rain-water, each edge of which is a centimetre: unit of weight.

From these four units all others are derived, according to the decimal scale.

Every system of weights and measures *must* have an *invariable unit* for its base; and every other unit of the entire system should be derived from it, according to a fixed law.

The French Government, in order to obtain an invariable unit, measured a degree of the arc of a meridian on the earth's surface ; and from this computed the length of the meridional arc from the equator to the pole. This length they divided into ten million equal parts, and then took one of these parts for the unit of length, and called it a METRE. The length of this metre is equal to 1 yard, 3 inches, and 37 hundredths of an inch, very nearly. Thus they obtained the length of the unit which is the base of the Metric System of Weights and Measures.

The next step was to fix the *law* by which the other units should be obtained from the base. The scale of tens was adopted.

PRONUNCIATION.

ME'TRE.	ARE.	LI'TRE.	GRAMME.
Mil'li-me-tre.		Mil'li-li-tre.	Mil'li-gramme.
Cen'ti-me-tre.	Cen'tiāre.	Cen'ti-li-tre.	Cen'ti-gramme.
Dec'i-me-tre.		Dec'i-li-tre.	Dec'i-gramme.
Dec'a-me-tre.		Dec'a-li-tre.	Dec'a-gramme.
Hec'to-me-tre.	Hec'tāre.	Hec'to-li-tre.	Hec'to-gramme.
Kil'o-me-tre.		Kil'o-li-tre.	Kil'o-gramme.
Myr'i-a-me-tre.		Myr'i-a-li-tre.	Myr'i-a-gramme.

The following are the Weights and Measures established by law in France, and now in general use:

Measures of Length.

Double decametre.
Decametre.
Demi-decametre.

Double metre.
Metre.
Demi-metre.

Double decimetre.
Decimetre.

Measures of Volume, for Grain.

Hectolitre.
Demi-hectolitre.

Double decalitre.
Decalitre.
Demi-decalitre.

Double litre.
Litre.
Demi-litre.

Double demi-litre.
Decilitre.
Demi-decilitre.

Measures of Volume, for Liquids.

Double litre.
Litre.

Demi-litre.
Double **decilitre.**
Decilitre.
Demi-decilitre.
Double centilitre.
Centilitre.

Weights in Iron.

Fifty kilogrammes.
Twenty kilogrammes.
Ten kilogrammes.
Five kilogrammes.
Double kilogramme.
Kilogramme.
Demi-kilogramme.
Double hectogramme.
Hectogramme.
Demi-hectogramme.

Weights in Copper.

Twenty kilogrammes.
Ten kilogrammes.
Five kilogrammes.
Double kilogramme.
Kilogramme.
Demi-kilogramme.

Double hectogramme.
Hectogramme.
Demi-hectogramme.

Double decagramme.

Decagramme.
Demi-decagramme.

Double gramme.
Gramme.
Demi-gramme.

Double decigramme.
Decigramme.
Demi-decigramme

Double centigramme.
Centigramme.
Demi-centigramme.

Double miligramme.
Miligramme.

It will be seen from the above tables, that the weights and measures in general use do not at all follow the decimal scale; for, in all Weights, and in all Measures of Volume, each decimal measure *has its double and its half*, while the *tables* are constructed entirely on the decimal scale. This discrepancy between the tables and the numbers in use must give rise to much confusion, and is a striking departure from the decimal system.

The entire Metric System, as now used, is given, that the reader may not be at the trouble of searching elsewhere for references in reading the report of the Committee, and the other more elaborate and more important documents which follow.

PART II.

REPORT OF THE COMMITTEE.

THE metric system of France had its origin in the stormy hours of the French revolution. In the year 1790, Prince de Talleyrand, then Bishop of Autun, introduced into the constituent assembly of France the proposition to establish a new system of weights and measures on the basis of a single and universal standard.

Two standards were considered:

1. The length of a pendulum which should vibrate seconds at a given point on the surface of the earth; and,

2. A given portion of the arc of a meridian.

It was finally agreed to adopt the one-ten-millionth part of the quadrant of the meridian passing through Barcelona and Dunkirk, as the universal standard.

This distance is called a metre. It is equal to 39.37 inches in length, and is the base of the metric, or French system. From this primitive base, or standard, every weight and measure is derived by the application of the decimal scale of tens.

The larger, or multiple units of the base, are designated by prefixing to the base the Greek numerals; and the smaller units, or sub-multiples, by prefixing to the base the Latin numerals.

In space, there are four kinds of quantity to be measured, viz., 1st, Distances; 2d, Surfaces; 3d, Volumes; and, 4th, Angles; hence, there must be four units of measure.

I. DISTANCE.—For this the unit is the metre, which is increased and diminished according to the scale of tens.

II. SURFACES.—For small surfaces, the square metre, and the squares constructed on its decimal divisions, are used. For the area of land, the base-unit is the square constructed on the deca-metre, called an *Are;* hence, an *are* contains 100 square metres, each of which is called a centi-are. In this measure, therefore, there are three units, viz.,

Centi-are = 1 square metre = 1.195985 square yards;
Are   = 119.6 sq. yards = 4 perches, nearly;
Hectare =   = 2.471 acres.

III. CUBIC MEASURE, or Measure of Volume:—

1. The cubic metre, which is used for the measurement of excavations and embankments. It is equal to 35.316 cubic feet.

2. The cubic metre is also used for the measurement of wood, and is then called a stere, which has one sub-multiple, called a deci-stere, and one multiple, called a deca-stere. Hence, for wood measure, there are three units:—deci-stere, stere, and deca-stere. The base-unit, the stere, slightly exceeds a quarter of a cord.

3. The litre is used for the measurement of all liquids and dry articles. It is the cube constructed on the deci-metre as an edge; and hence, is one-thousandth part of the cubic metre, equal to one wine quart, very nearly.

IV. For Angles.–Although the metre is derived from the arc of a meridian, it is a singular fact that neither it, nor any multiple or sub-multiples of it, is used in the measurement of circles or angles. In this department of measurement alone, the old unit and its subdivisions, are preserved.

Weight.—The unit of weight is the weight of a cube of water constructed on the centimetre as an edge; and is called a *gramme*. It is equal to about 15.32 grains Troy. The kilogramme, equal to 1000 grains = $2\frac{1}{2}$ lbs., is the unit for larger weights.

Before examining this system of weights and measures, a few remarks seem necessary in regard to the nature and uses of numbers.

Every number is derived from a fixed unit, or base, either by the process of augmentation or division; and that base is invariably called one, or the unit of the number. Thus, five comes from 1, by taking it 5 times; and one-fifth comes from 1, by dividing 1 into five equal parts. To apprehend distinctly the signification of a number, two things are necessary:

1st. To apprehend, distinctly, the *unit* of the number; and, 2dly. To comprehend, clearly, *how many times* that unit is taken.

If the base-unit is too large, two difficulties are encountered: 1st. The difficulty of apprehending its exact value; and, 2dly. The smaller numbers must be expressed fractionally. Thus, if our smallest unit of length were the yard, it would not be so easy to impress the mind of a child with its exact value, as it is to give the idea of a foot; and all distances less than a yard would be expressed fractionally. Thus, the foot would be one-third of the yard, and the inch one-thirty-sixth of the yard, and these would be very indistinct impressions compared with our present system, in which the foot is the base-unit, and the inch, one-twelfth of it.

Again, if the base-unit be too small, it must be repeated many times in order to express large quantities; and the mind finds much difficulty in apprehending a number in which the unit is taken many times. Thus, we comprehend very easily, 1 mile, when we have reached it through the units, 1 foot, 1 yard, 1 rod, and 1 furlong, each of which being expressed in small multiples of the preceding unit, is easily apprehended. But if we say a mile is 1760 yards, we have to reflect a little before we comprehend the distance, as the number arises from taking the base-unit, 1 yard, 1760 times; and if we call the mile 5280 feet, we apprehend it still less distinctly; and the mind is hardly willing to make an analysis, when we say, the mile is 63,360 inches.

Now, the French system adopts the metre as the only unit of distance. No other unit is permitted. Hence, all distances greater than the metre are expressed in multiples of the metre—and all distances less than the metre, in decimals of it. The distance to the sun and the diameter of the spider's web are both measured by the same unit; which is quite too large to measure the one, and quite too small for the measure of the other.

These are some of the difficulties resulting from a single unit of length. It is of the first importance, in the construction of a system of numbers, to have all the different units flow directly from the *base-unit* and from each other, without the intervention of **fractions.** In our system, **the** square foot and the cubic **foot, the square** yard and the cubic yard, are but squares **and cubes** constructed on the units of length.

For **the** measurement of land, we reach the acre **through the** perch and rood; and the acre is **a unit of** convenient size for large areas. In the French system, **the** smallest **surface,** expressed **without** the aid of fractions, is **the** square metre, **equal to about 10** square feet; the next smaller unit is the square on **the decimetre,** which is equal to one-hundredth of the square **on the metre, and so on,** for units still less. Thus, in this system, all the units of surface, less than the square on the metre, are obtained by squaring a decimal part of the metre; and the units thus obtained, besides their fractional **origin, have** no names, **and are connected, not by the scale of tens, but** by the scale **of** 100.

The principal **unit for the measurement of land** is the **square** constructed **on the decametre, which is** called an *Are,* and which is equal to about four perches, or, the one-fortieth of our acre. One hundred of these make the hectare—so that the units for **the** measurement of land are—

The centiare, or square metre;
The are   = 4 perches, nearly; and
The hectare = about 2½ acres.

The adoption, therefore, of a single unit for the measurement **of length** carries with it the necessity of constructing other base-units, either on multiples or on fractional parts of the unit of length; and also, of a change of scale, in expressing the relations of the derived units.

Thus, for land measure, the *are* is a square constructed on the decametre, and the ratio of the units, to each other, is 100, instead of 10.

The *litre*, which is the unit of measure for dry articles and liquids, is the cube constructed on the decimetre as an edge, and is equal to the one-thousandth part of the cubic metre, to which it must be referred before we can apprehend its value.

Standard Units of Length now in use.

The smallest unit named in our tables is the barley-corn, three of which, laid in a row, make an inch. The inch, also, has a symbol in the distance from the end of the middle finger to the first joint, and twelve such distances make a foot. The foot is the unit of length with which we are most familiar, and was undoubtedly adopted as a standard of measure from that part of the human body from which it takes its name. The average measure of the naked foot is but the fraction of an inch below the standard; and in its ordinary covering approximates very nearly to it. The foot has, also, another type in the human body—the average distance across the hands when the extremity of one thumb is placed at the upper extremity of the nail of the other.

The first measure named in sacred history is the cubit—the average distance from the elbow to the extremity of the middle finger. The yard, also, has its representative in the human body—being the average distance from the centre of the lips to the extremity of the middle finger, when the arm is extended. The fathom is the distance between the extremities of the middle fingers, when both arms are extended.

The pace, or long step, also corresponds to the yard, and is the natural unit for all itinerary distances. Thus, all the elementary units of measure, except the barleycorn, are derived, directly, from the human body, and every individual carries about him an approximate standard.

It has been urged against the continuance of our present system, that its units are derived from the varying and uncertain standards of barleycorns and the feet of men; while, in fact, these are only the sensible objects from which the exact units are derived, and to which they are referred for the purpose of impressing their values on uninstructed minds. Our standard of length is the standard yard of the British empire, and is symbolized by a metallic bar, made by Bird in 1760, when at the temperature of 62° Fahrenheit. By Act of Parliament in 1824, it was provided that in case of total destruction or loss of the standard and all its authentic copies and facsimiles, that its length be considered 36 inches, *such*, that 39.13929 of them are equal to the length of a pendulum vibrating seconds, *in vacuo*, at the sea-level in the latitude of London; and this, we shall see, is precisely the means which the French have adopted to replace their standard, in case of accident. Hence, the two systems rest on a common principle.

Question to be considered.

The question is now under consideration, both in this country and in Europe, whether the French system of weights and measures shall become the universal standard? Whether the metre shall be adopted as the only standard of length—whether we can adopt the *Are* as the base and only unit of all land measurements—the *Litre* for all volumes—and the *Gramme* as the unit of all weights? These questions have already been carefully and fully considered.

On the 14th of December, 1819, the House of Representatives passed the following resolution, viz., "*Resolved—* That the Secretary of State be, and he is hereby, directed to report to this House a statement relative to the regulations and standards for weights and measures in the several States, and relative to proceedings in foreign countries, for establishing uniformity in weights and measures, together with such a plan for fixing the standard of weights and measures for the United States as he might deem most proper for their adoption."

In obedience to this resolution, on the 22d of February, 1821, John Quincy Adams, then Secretary of State, made a very full and able report. The report is a complete and careful analysis of the whole subject, by a mind full of knowledge, and looking at it from the standpoint of science and statesmanship. The subject has also been fully examined by Sir John Herschel, in his lecture on the yard, the pendulum, and the metre, considered as a standard.

It may be regarded as fortunate for the interests of knowledge, that a subject so important and so full of interest to a hundred millions of people should have been examined and passed upon by two such minds; and that they should have left to posterity the rich treasures of their labors.

The decree of the Convention of August, 1793, established, in France, all the principles of the new system. Its denominations were entirely different from those before

in use, and from any which have since been adopted.
For many years thereafter, there was continued change
and conflicting legislation. Finally, on the 12th of February, 1812, an imperial decree was issued which presented
the whole subject under a new aspect. It, in fact, restored
the old system practically, and retained the new one only
in name. The following is an analysis of it, given by Mr.
Adams:

"It cannot escape observation that this decree and
explanatory ordinance engrafted upon the legal system an
entirely new system, founded upon different, and in many
important respects, opposite principles, so that the result
up to this time (1821) of the most stupendous and systematic effort ever made by a nation to introduce uniformity
in weights and measures, has been a conflict between four
distinct systems—

"1. That which existed before the revolution;

"2. The temporary system established by the law of 1st
of August, 1793;

"3. The definitive system, established by the law of December 10th, 1799; and

"4. The *usual* system, permitted by the decree of 12th
February, 1812.

"This last decree is a compromise between philosophical
theory and inveterate popular habits.

"Retaining the principle of decimal multiplication and
division for the legal system, it abandons them entirely in
the weights and measures which it allows the people to use.
Instead of the metre and its decimals, it gives the people
a toise of six feet, an une of three feet, and a thumb of
twelve lines. And these measures, instead of divisions,

exclusively decimal, are divisible into halves, thirds, quarters, sixths, eighths, twelfths, and sixteenths.

"Instead of a decimal kilogramme, it gives them a pound of sixteen ounces, an ounce of eight gros, and a gros of twenty-two grains.

"The measures of capacity, wet and dry, have the same indulgence: and while the standard weight and measure are deposited in the national archives, the people have restored to them, for use, all the *names* and divisions of their ancient weigths and measures, though not the same things. For the toise, which is twice the length of the metre, is not the old toise; the foot, which is the third part of the metré, is not the pied de roi; but both are longer measures. The half-kilogramme, which is a pound, is not the ancient mark-weight pound; nor are the toisseau or litre those of ancient times; they are all, respectively, approximations of them.

If the existing system and practice terminated here, it would be far from having attained the ideal perfection of uniformity; but it is believed that, for a multitude of purposes, with this double and complicated system, there is yet a very extensive remnant in use of that which prevailed before the Revolution.

"The changes which have forced themselves upon the new system, under the attempts to reduce it to practice, should serve as admonitions to correct the errors of theory; but not operate as discouragements to the pursuit of the principal object, *uniformity.* The French metrology, in the ardent and exclusive search for an universal standard from nature, seems to have viewed the subject too much with reference to the nature of things, and not enough to the nature of man. Its authors do not appear to have considered, in all the bearings of the system, the proportions dictated by nature between the physical organization of man and the *unit* of his weights and measures. The standard taken from the admeasurement of the earth, has no reference to the admeasurement and powers of the human body. The metre is a length of forty inches, nearly; and by applying to it, exclusively, the principle of decimal divisions, no measure corresponding to the ancient *foot* was provided. A unit of that denomination, though of slightly varied differences of length, was in universal use among all civilized nations; and the type of it is found in the dimensions of the human body. Perhaps for half of the occasions which arise in the life of every individual for the use of a linear measure, the instrument to suit his purposes must be portable, and fit to be carried.in his pocket.

"Neither the metre, the half-metre, nor the decimetre, are suited to that purpose. The half-metre corresponds, indeed, with the ancient cubit; but, perhaps, one of the causes which have everywhere, since the time of the Greeks, substituted the foot in the place of the cubit, has been the superior convenience of the shorter measure. Besides which, the cubit being the unit, the half-cubit might serve the purposes of the foot; but the metre, divisible only by two and ten, gave no measure practically corresponding to the foot, whatever. It appears to have been considered that decimal arithmetic, although affording great facilities for the computation of numbers, is not equally well suited for the division of material substances. A glance of the eye is sufficient to divide material substances into successive halves, fourths, eighths, and six-teenths. A slight attention will give thirds, sixths, and twelfths. But divisions into fifth and tenth parts, are among the most difficult that can be performed without the aid of calculation.

"Among all its conveniences, the decimal division has the great disadvantage of being itself divisible only by the numbers two and five. The duodecimal division, divisible by two, three, four, and six, would offer so many advantages over it, that while the French theory was in contemplation, the question was discussed, whether the reformation of weights and measures should not be extended to the system of arithmetic itself, and whether the number twelve should not be substituted for ten, as the term of the periodical return to the unit. Since the establishment of the French system, this idea has been reproduced by philosophic critics as an objection against it; and Dalambre, in the third volume of the Base du Système Matrique, has considered it, and assigned the reasons for which it had been rejected. He admits, to the full extent, the advantages of a duodecimal over a decimal arithmetic, but alleges the difficulty of effecting the reformation, as the decisive reason against attempting it."

The decree of February 12th, 1812, was the triumph of the popular will over the acts of the government. It practically abolished the new system, and re-established the old one. It arrested those attempts at change which had caused so much trouble and perplexity, and brought again into use old names and old things, with which the people were familiar.

But the conflict between legislation and the popular

will, between what the law required and what the law permitted, did not end here; it was continued for a quarter of a century, when the whole subject was finally disposed of by the strong arm of power.

A law of July 4th, 1837, repealed the decree of February 12th, 1812, and ordered the exclusive use of the decimal metric system. A royal decree of April the 17th, 1839, regulated the relations of weights and measures. A decree of July the 16th, 1839, determined the denominations, form, and dimensions of all instruments and measures, for trade and use. The metric system is now obligatory and is exclusively used in the French empire, and has been so since 1840, the time named for the entire discontinuance of all other systems.

The system, however, as now established, is not exclusively decimal. The measures of distance, sanctioned by the laws and decrees of 1837 and 1839, are, the metre, the demi-metre, the double decimetre, and the decimetre; and in the ascending scale, the double metre, the demi-decametre, the decametre, and the double decametre:— eight in all.

For the measurement of land, there are but three units —the centiare, or square metre; the *are*, and the hectare; and in this measure, in passing from one unit to another, the scale is 100. In the denominations of volume and weight, each unit has a *half* and a *double;* so that, in fact, there are three times as many units in use, in these denominations, as there are in the tables; and this concession was necessary, for the half and the double would not be excluded.

The departure from the decimal system has greatly

multiplied the number of units, so that now there are eight units of linear measure, three of square measure, three of land measure, fourteen of liquid and dry measure, six of solid measure (including three for wood), and twenty-three in weight—making fifty-seven in all; while the tables of the system contain but twenty-eight units—the halves and the doubles being entirely omitted. This discrepancy between the tables and the units in general use is, in fact, a fatal difference between theory and practice, and must lead to complexity and embarrassment.

The friends of the metric system urge its universal adoption, for three reasons:—

1st. That it is derived from the metre, which they declare to be a *fixed*, and natural unit or standard;

2d. Because the decimal scale is employed in forming the multiples and submultiples of the base-unit;

3d. Because of the simplicity and comprehensiveness of its nomenclature.

That the metre, as determined by the French measurement, is neither a true nor an accepted standard, is a fact well known to all men of science; as the following extract from the lecture of Sir John Herschel, the greatest of living astronomers, fully shows. He says:—

"Let us now see how far the French metre, as it stands, fulfils the requirements of scientific and ideal perfection. It professes to be the 10,000,000th part of the quadrant of the meridian passing through France from Dunkirk to Formentera, and is, therefore, scientifically speaking, a local and national, and not a universal measure. The earth's equator is not a perfect circle, but slightly elliptical, and the meridians of places differing in longitude are, therefore, not all of the same length. The difference, however, is so trifling (the ellipticity of its equator being not more than a thirtieth part of that of its meridian), that, to raise an objection against the practical reception of the metre, either *per se*, or as a substitute for the yard, on this score, would savor of hypercriticism. A more serious objection is the choice made of the circumference of the meridional or generating ellipse of the terrestrial spheroid, in preference to its axis of revolution. This is a blemish on the very face of the system—a sin against geometrical simplicity. Still, were the length of the metre, as determined by the French geometers, rigorously exact, or correct within limits which the much more extensive measurements of meridian arcs, since made, elsewhere than in France, have proved to be attainable, this would be only a matter of regret, and could hardly, of itself, be drawn into an argument for its rejection. But this is far from being really the case. The metre, as represented by the material standard, adopted as its representative, is too short by a sensible and measurable quantity, though one which certainly might be easily corrected."

To this testimony must be added that of Mr. Airy, the astronomer having charge of the Greenwich Observatory, and M. Schubert, a Russian astronomer of great eminence, pointing out, specifically, the extent of the error, and giving reasons why the metre should not be accepted as a standard. Without stopping to examine and weigh these reasons, it is enough for our present purpose to know, that the science of the world has not accepted the one-ten-millionth part of the quarter meridian, as having a *fixed value*, represented by the metre, and that the ablest minds in England will probably **not so accept it.**

Furthermore, as stated by Sir John Herschel, "The report of the French commissioners, in 1798, which led to the enactment of the metrical system, is careful to state that in the event of the total loss or destruction of all material representations of the metre, its value would be easily recoverable from a numerically specified relation between its length and that of the pendulum vibrating seconds at Paris, which had been determined with great accuracy by Borda, one of the commissioners. So that, practically speaking, in the event of the total destruction, by political convulsions, of every authentic yard and metre (supposing every written record of an existing knowledge to survive them), the metre would have been recovered, not by the laborious and costly process of remeasuring the French meridian arc, but by the infinitely more summary one of a precise repetition of Borda's experiments, and the exact reapplication of all his corrections and reductions:" so that the two standards, the English and the French, rest, essentially, on the same basis.

2. In regard to the simple use of the decimal scale, we have already shown that in most of the weights and measures, each unit has a half and a double, where, of course, the scale of connection is two, not ten; and this having been adopted from necessity, after the adoption of the system itself, one-half of the units in common use are not in the tables at all—so that the pupil, after having learned his table-book at school, has a new set of units to learn in practical life.

Besides, one of the four great divisions of measure, viz., the arcs of circles, including all that belongs to the earth in geography, and all that belongs to the heavens, as examined and measured by the science of astronomy, the attempt to apply the decimal system has utterly failed, and is now entirely abandoned.

3. The nomenclature, by taking the four base-units, the METRE, the ARE, the LITRE, and the GRAMME, and applying to them the Greek numerals for the multiples of the unit, and the Latin numerals for the sub-multiples, has been thought to simplify, very much, the language of the entire system ; and so it undoubtedly does to the mind and apprehension of a scholar. But it must be remembered that the tables of arithmetic are learned and used by thousands who will never know the difference between deca and deci, between hecto and centi, as derived from the Greek and Latin ; and hence, deca-metre, hecto-metre, deci-metre, and centi-metre will have to be learned as separate words. Now, all teachers have experienced the great difficulty in communicating ideas differing widely from each other, when the language used is nearly the same ; and a child, knowing nothing of the Greek and Latin, would find greater difficulty in distinguishing between deca-metre and deci-metre, between hecto-metre and centi-metre, than he would if the things were called by entirely different names. If he do not *know the law*, the similarity and resemblance of the words are great hinderances. These remarks are alike applicable to the nomenclature of the entire system.

Question Soon to be Decided.

It is generally known that Congress, in the year 1866,

passed a law permitting the use of the metric system, and directed certain of its weights and measures to be sent to offices having a close connection with foreign countries. The question is now being discussed, and must soon be passed upon, whether this system shall be adopted in the United States. Having explained very briefly the nature of the system, and something of its history, the convocation will readily understand the reasons which have governed the committee in the conclusions to which they have arrived.

1. If the metric system be introduced, it must supercede all present systems. We must adopt the system as a whole, and exclude every other. The history of its introduction and early use in France, clearly prove this. The French people struggled against it for twenty years; and finally, practically overthrew it with the concurrence and sanction of the great Napoleon, in the palmiest days of the empire. Then followed twenty-five years of confusion, under different and conflicting systems, when a new revolution so strengthened the power of the government that the metric system was re-established in 1837, and the use of any other weight or measure made a penal offence. That the conflict will be fierce in this country, where the people are freer and less habituated to blind obedience to imperial edicts, cannot be doubted; nor will the fact, that the system comes from a foreign country, whose language and institutions are alike unknown to us, be without its influence.

2. An unwillingness to abandon what has been tried and known, for what is untried and unknown, appears to be a law of the human mind. It is a species of intellectual

inertia, corresponding to the inertia of matter that pervades all bodies. We have, in the history of our own currency, a striking illustration of this principle.

One of the first acts of Congress, after the establishment of the government, was to define and fix the currency. The unit selected was the best possible—the dollar, and the divisions of it, according to the decimal scale, gave promise of an early adoption, and an entire uniformity. But nearly eighty years have now elasped, and the desired uniformity is not attained; for, in many places accounts are still kept in the old currency. Such are the fruits of the force of habit, and of tenacity to established systems.

3. The committee do not believe that the opposition to the metric system, in France, arose solely from an unwillingness to change names and forms, or from a blind attachment to old usages—though this doubtless did, and always will, exercise a strong influence.

Their old system of weights and measures sprung up from the common necessities of mankind. All the units of measure, with which the nation was familiar, were found, very nearly, in the parts of the human body. The relation between those units had been established by the people themselves, either from computation, or measurement, or the exchange of their commodities. A system having such an origin was more likely to meet the wants of a people than one made amid the turbulence of a revolution, by a committee of learned professors.

4. The committee has already referred to some of the objections to any system of numbers where the multiples and submultiples are all formed from a single unit, as a

base. We will now consider other objections from the standpoint of the shoolroom.

Educational View.

Every teacher knows that the first step in a course of arithmetical instruction is, to impress the pupil with a distinct and full apprehension of the unit of number, whether that unit be abstract or denominate.

What shall be the unit for distance, and whether there shall be more than one, are important questions.

The entire system of instruction in mathematical science —all weights and measures, and, to a certain extent, the mechanic arts, are concerned in these questions. It is the opinion of the committee, that the metre is too large for a base-unit:

1st. Because it is not easy to give a young and uninstructed mind a distinct apprehension of it; and

2dly. Because there are many things to be measured, in the common affairs of life, less than the metre, and these must all be expressed in fractions of that unit.

The committee are also of opinion, that other units, besides the base-unit, should be used as secondary bases of collections of numbers.

Taking the foot as the unit of distance, in our own system, we measure all the distances less than it by its divisions into twelfths, and all distances greater may be measured by the foot unit. But the apprehension becomes dim as the numbers grow large; and young minds, in computation, must be trained in small numbers. Hence, we take a second unit, 1 yard = 3 feet; and instead of saying 300 feet, we say 100 yards. We then take the rod, the furlong, and the mile, as units, in succession, and thus

avoid the use of large numbers, for small distances, by conducting the mind, gradually, through a series of ascending
units, instead of expressing all numbers, great and small,
by the same unit.

What would follow.

Let us suppose the metric system to be adopted by law,
and every other system excluded—for without such exclusion, the whole thing would be a perplexity and a farce.
What follows: we have blotted out, from the mind of the
nation, the foot and all knowledge of every measure into
which it enters, as a unit. We have expunged the yard,
used in connection with the arm, more or less in every
family; and the pace, the unit and guide of the farmer,
for an approximate measure, that will not supply the
place of either.

Every lot of ground 25 feet front, by 100 feet deep, must
be described as follows: 7 metres, 6 decimetres, and 2
centimetres front, by 30 metres, 4 decimetres, and 8 centimetres deep. Thus, the description of every such lot will
require three different units and six words, instead of one
unit and two words. In all conveyances and descriptions
of land, the translation from one language to the other
would occasion great trouble and difficulty.

The old familiar mile, of 1760 paces, is also gone, and
the distance from Albany to New York, one hundred and
forty-five miles, will be known to us, if known at all, as
229,680 metres.* Let us see how we shall recognize the
earth, in its new dimensions. Its diameter, instead of
eight thousand miles, in round numbers, will be 12,672,000

---

* In these calculations, the metre is taken equal to 40 inches.

metres; and its circumference, about 39,810,355 metres and 2 decimetres. Is not the difficulty of apprehending and comparing large distances greatly increased, by expressing them in small units and large numbers?

The cubic foot, known wherever the English language is spoken, as the simplest unit for the measurement of volume, is also gone, and in the twilight of its existence we grope about for a substitute. We find it in the cubic metre, about thirty-five times as great as the cubic foot—a value quite too large to be apprehended by children, and altogether too large for a base-unit. Or, we find it in the litre, the cube constructed on the decimetre as an edge—equal to one-thousandth part of the cubic metre, or to the one-four-hundredth part of the cubic foot; a unit quite too small for common measurements. Into this unit merges one gill, one quart, one gallon, one peck, one half-bushel, and one bushel. They all disappear, and we measure in the litre, and its multiples and submultiples. The acre is also gone, with its submultiples, the rood and perch, and all its linear dimensions. For it, we have the *are*, about one-fortieth of its value: so that a quarter-section of land of 160 acres, will be known to us as containing 6,400 ares. Since the commencement of the present century, the public lands have been surveyed and laid out in townships six miles square, each containing, of course, thirty-six square miles, or 23,040 acres. The side of each township, by the new system, would contain 9,504 metres (instead of six miles), and its area, 921,600 *ares*. All the lands, from the Ohio River to the Pacific Ocean, have been surveyed, deeded, and recorded in the units of the square mile and the acre. What will be the labor and the confusion of translating every deed and record into the language of the metre and the are? We should scarcely know our own farms by their new names.

Effect on Weights.

We come next to the unit of weights. In our own system, it is derived directly, and in whole numbers, from the cubic foot. A cubic foot is supposed to be filled with distilled rain-water, and one-thousandth part of this weight is taken as the unit—the ounce avoirdupois. This being too small for the measure of ordinary weights, its multiple by 16, or the pound avoirdupois, is taken as the unit of weight; and from this come the multiples—quarters, hundreds, and tons.

In the Metric System, a cube, described on the centimetre as an edge, is filled with water at the freezing-point, and this weight is taken as the unit, and is called a *gramme*. It is equal in value to about the 454th part of one pound, avoirdupois. This small unit is made the *base* of the entire system of weights. Hence, the weights of all common articles are expressed in very large numbers. For example, a piece of beef, for dinner, which we designate by the modest number, 14 pounds, would have its weight expressed by 6,356 grammes; or by 6 kilogrammes, 3 hectogrammes, 5 decagrammes, and 6 grammes; whilst a ton of coal of 20 hundred pounds would be read, 908,000 grammes; or 1 quintal and 8 kilogrammes. As a general rule, all the readings of numbers are made in the lowest unit. In weights, therefore, as in the other denominations, corresponding difficulties arise from the smallness of the base-unit, and from using but a single base. If the introduction of the Metric System produced only a change in the *names* of the units, leaving their values the same; or, if it altered the values only, preserving their names, the difficulties would be comparatively small. But, unfortunately, we must change both ideas and words—the foundations of systems and the language by means of which these systems are developed and made known. These double changes, made at the same time, are very serious.

Changes in Values and Prices.

We must not forget that prices and currency are dependent upon, and necessarily adjust themselves to, weights and measures; and that all our ideas of cost and value are fixed with reference to our present units. The adoption of the Metric System, therefore, would carry with it an entire change in the money values of all articles of commerce and manufactures, and of all agricultural productions; for these values would have to be readjusted to the new units, and to be expressed in the new language.

Consequences of Making the Proposed Change.

1. It would strike out from the English language every word and phrase and sentence used in connection with our present units of weights and measures, and would impose the necessity of learning a new language for the one now in use:

2. It would blot out from the knowledge of the nation all apprehensions of distance, and area, and volume, acquired through the present units, and would render necessary the acquirement of similar knowledge by less convenient units, having different relations to each other, and expressed in a new and unknown language:

3. It would extinguish all knowledge of money values, now so familiar to the entire population in their daily purchases, and sales, and barters, for those values are all adjusted with reference to the units of weights and measures: and

4. It would change the records of our entire landed property, requiring them all to be translated into a new and foreign language. Should all this be done merely to change one standard from 36 inches to 39.37 inches, when both standards are determined, substantially, in the same manner?

The convenience, the almost absolute necessity, of a common unit of coin among commercial nations, is too apparent for argument, and this urgent necessity has suggested common units of weights and measures. But these, although highly desirable, are by no means so important. The sale and purchase of land very seldom reaches to foreign countries; and, hence, there is little need of a common unit. In weights and in measures other than land, a common unit is certainly desirable, but not so much so as in currency. The practical question presented is, whether the advantages resulting from the changes suggested, would be a compensation for the trouble and perplexity that must arise in making them?

But even if the changes, on the whole, be deemed desirable, it is obvious that they should only be made at the same time and in conjunction with their adoption by the English government.

With the English people we have a common origin, a common history, a common literature, and a common language. Would it not be most unwise to interrupt our commercial relations with such a people, by the introduction of a new and foreign language, to them and to ourselves, into trade and commerce?

Decimal System.

The Metric System has been recommended to our adoption because of its decimal scale; and the twin ideas of a common unit for all nations, and the decimal multiples and submultiples of that unit for all numbers, are certainly most attractive. But Bacon has proved to the world that systems developed from abstract theories are not as likely to prove useful and satisfactory, or to be in accordance with true philosophy, as those which are founded on facts derived from observation, experiment, and experience. For the purpose of *calculation alone*, the decimal system is greatly superior to every other; but for sensible objects, which are daily measured and handled, the French themselves have departed from it by introducing the half and the double, for most of the units, as may be seen from the tables.

In a system of instruction, it is as necessary to preserve the regular order of the *fractional units,* when we divide the unit, as it is to preserve the order of the numbers, when we collect or aggregate it. In whole numbers, we reach ten only through a knowledge of the preceding numbers from 1 to 10. So of one-tenth, we apprehend it only when we have divided the unit into two equal parts, into three, into four, into five, and so on up to ten. The *fractional units,* one-half, one-third, one-fourth, one-fifth, etc., must each and all be clearly apprehended before the mind can grasp, as a crystallized idea, the fractional unit one-tenth. Hence, no system of instruction can dispense

with the divisions of the unit into any number of equal
parts, nor can positive legislation effect it. So in material
things. The simplest idea of division is to halve the thing
to be divided, then to divide it into thirds, then into
fourths, and so on for any division whatever.

It was a most fortunate circumstance that Congress, in
the early history of the Government, established a decimal
currency. It is much to be regretted that there is any
other. This Act, alone, gave us the decimal system, for all
money values and for all computations of price and cost;
and these embrace by far the larger portion of our calculations. It did even more. It turned attention to the
beauty, the simplicity, and the harmony of the system itself, and we are rapidly approaching to its exclusive use
in all cases of computation. The chain, used in surveying
land, is of such length that 10 square chains make an
acre. It is divided into 100 links, so that chains and links
multiplied by chains and links give, in the product, acres
and decimals of an acre. Here, in computation, the decimal system is perfectly applied; and surveyors now omit
the rood and perch, and write their final results in acres
and decimals of the acre. This tends to simplicity, and
introduces no confusion. So in levelling—the staves are
now divided, and the field-notes are kept in feet and decimals of the foot, and the computations are made entirely
in these units. In field-notes, where yards are used, the
fractions of the yards are generally recorded in the decimal
language. Thus, we are gradually approximating to the
decimal system, and if not embarrassed by legislative enactments, we shall soon reach it, in every case to which it
can be advantageously applied.

Conclusions.

The committee, for the reasons above given, have not been able to see how any system of weights and measures can be an acceptable substitute for the one now in use, unless it makes some provision for retaining the unit, one foot. All our knowledge of distances, the yard, the rod, the furlong, the mile, the league, come from it. The square rod or perch, the rood, the acre, are also derived from it. Can we change the survey of an entire continent, with the description of every piece of land upon it, from the unit, one Acre, to the unit, one Are, forty times less? Can we change, without great confusion, the units of volume, the cubic foot and the cubic yard, so familiar to every school-boy? and, above all, can we change our unit of weight, the pound avoirdupois, which is equal in weight to sixteen of the one-thousand equal parts of a cubic foot of rain-water? It seems to the committee that we must retain the following units: viz.,

1 foot in length,	1 square foot,	1 cubic foot;
1 yard,	1 square yard,	1 cubic yard;
1 rod,	1 square rod, or perch,	1 acre;
1 mile,	1 square mile, or	640 acres;
1 quart, (liquid)	1 gallon,	1 barrel;
1 quart, (dry)	1 peck,	1 bushel;
1 pound.	1 hundred,	1 ton.

Can we abandon, as a mere question of language, these short, sharp Saxon words, for their equivalents expressed in a foreign language? Besides, the foreign language which we introduce has no exact equivalents to these words, which have almost become things, and which now form a part of the mind and knowledge of every people which speak the English tongue, or are connected with American commerce.

Weights.

On the subject of weights the duty of the committee seemed both plain and imperative. We have the Avoirdupois, the Troy, and the Apothecaries' weight. The pound in the Avoirdupois weight, differs from the pound in the other two, and in each the ounce is differently divided.

The attention of the committee was first directed to the inquiry—how can these three systems of weights, differing apparently so widely, be reduced to a single system with the least possible change and inconvenience?

The Avoirdupois weight being much more used than either, or both of the others, they of course should be assimilated to it; and the Apothecaries' weight, being used more than the Troy weight, it is less inconvenient to change the latter than the former.

In analyzing these weights, it is found that the ounce, in the Apothecaries' weight, and the ounce in the weight Troy, are identical; and that each exceeds the ounce Avoirdupois by its eighty-three-thousandth part, very nearly; hence, if the ounce Troy, or the Apothecaries' ounce, be diminished by its eighty-three-thousandth part, the result will be the ounce Avoirdupois, or the one-thousandth part of the weight of a cubic foot of distilled rain-water, and then, *these three weights will have a common unit.* As the Avoirdupois is now mainly used for all weights greater than an ounce, and the others for weights less than the ounce, the three can be united on that common base, with very little inconvenience. The new Avoirdupois weight would then be as follows:

20 grains   make 1 scruple,
3 scruples   make 1 dram,
8 drams   make 1 ounce,
16 ounces   make 1 pound,
25 pounds   make 1 quarter,
4 quarters make 1 hundred, and
20 hundreds make 1 ton.

This would preserve, in the Apothecaries' weight, all the units including and below the ounce, with this only difference, that each would be eighty-six thousandths less than before, while the change in the value of the pound and in the units of higher denominations, would be attended with little or no embarrassment, as the Avoirdupois weight is now chiefly used in all wholesale transactions.

In the Troy weight, the principal unit, the ounce, would be preserved in name, with the diminution of only eighty-six thousandths of its value; the ounce would still contain 480 grains, as now, and the only change would be, the substitution of the scruple and dram, for the pennyweight. These slight changes would perfectly harmonize the three systems of discordant weights.

Coins and Currency.

The questions connected with coins and currency have seemed to the committee very plain and simple.

We are indebted to the wisdom of Congress, at the very birth of the Government, for a system of coins and currency perfect in its general outline, and admirable in all its details. The unit, one dollar, is a better unit than the franc, which is too small, and better than the pound sterling, which is too large. Its decimal multiples and submultiples are in accord with the decimal system, the most perfect for computation which the world has ever known.

It is, however, a subject of equal surprise and regret, that a system of currency so perfect, and having the force and the authority of law, should have found its way so slowly into general use.

About eighty years have now elapsed since it became the legal and only recognized currency; and yet, in all our standard arithmetics we find New England, Virginia, Kentucky, and Tennessee currency, in which six shillings are reckoned in the dollar; New York, Ohio, and North Carolina currency, in which eight shillings are reckoned in the dollar; New Jersey, Pennsylvania, Delaware, and Maryland currency, where 7s. 6d. are reckoned in the dollar; and South Carolina and Georgia, where 4s. 8d. are reckoned to the dollar; and if our information be correct, accounts are still kept, in many of the States, to a greater or less extent, in these old continental currencies.

The first step toward uniformity at home, should be the entire elimination from our elementary arithmetics of all these antiquated currencies. To consume the time of the young, at school, in teaching old systems long superseded by better ones, is little short of a crime.

In regard to foreign coinage, the committee have great satisfaction in stating that a near approximation to a common unit of value has already been reached, and that an agreement on a common unit would seem to be easily attainable.

The five-franc piece is so nearly equal in value to our dollar, and five of our dollars so nearly equal to the English pound sterling, that a very slight change in value would make the dollar and the pound multiples of the franc, and thus reduce the currency of the three great commercial nations to a common unit. Two of them, France and the United States, have already adopted the decimal division; and the report of Sir John Bowring on the decimalization of the English currency, so clearly proves the great value of the decimal system over every other, that there can be no reasonable doubt of its early adoption.

The committee have embodied the substance of their report in a series of resolutions, which they now beg leave to submit:

1. *Resolved.* That the subject of changing our entire system of weights and measures and substituting therefor the Metric System of France, is too grave and too important to be acted upon without a very full and careful examination of all its bearings and all its consequences.

2. *Resolved.* That the committee on coinage, weights, and measures, be requested to publish their report to this convocation, with such additions as they may deem necessary, in connection with the report of John Quincy Adams, on weights and measures, made to the House of Representatives, in the year 1821, and the lecture of Sir John Herschel, on the Yard, the Pendulum, and the Metre, to the end that the whole subject may be more fully discussed, considered and understood.

3. *Resolved.* That this convocation do recommend to all teachers and to all others interested in the establishment of uniform standards throughout the world, to give special attention and study to this subject, now engrossing

public attention, that it may be finally disposed of wisely, and for the common interest of all nations.

4. *Resolved.* That this convocation do hereby express its conviction that such changes should be made in the values of the franc, the dollar, and the English pound sterling, that five francs be exactly equal in value to one dollar, and five dollars exactly equal in value to one pound sterling.

5. *Resolved.* That as a means of reaching uniformity of currency, it be, and it is hereby, earnestly recommended to all authors and publishers of elementary arithmetics, to exclude from future editions every currency not recognized and established by law.

6. *Resolved.* That the committee on coins, weights, and measures be, and they are hereby, authorized to ask the attention of the government, and of all associations for the advancement of science and knowledge, to the expediency of changing the value of the ounce Troy, and thus substituting a single weight for the three now in use.

7. *Resolved.* That the committee on coins, weights, and measures be, and they are hereby, authorized to take such steps, by correspondence or otherwise, as will in their judgment be most likely to give effect to the above resolutions.

The above resolutions were adopted unanimously.

PART III.

REPORT OF JOHN QUINCY ADAMS.

THE SECRETARY OF STATE who, by a resolution of the House of Representatives of the 14th of December, 1819, was directed to report to the House "a statement relative to the regulations and standards for weights and measures in the several States, and relative to proceedings in foreign countries, for establishing uniformity in weights and measures, together with such a plan for fixing the standard of weights and measures for the United States as he might deem most proper for their adoption," respectfully submits to the House the following

REPORT:

The resolution of the House embraces three distinct objects of attention, which it is proposed to consider in the following order:

1. The proceedings in foreign countries for establishing uniformity in weights and measures.

2. The regulations and standards for weights and measures in the several States of the Union.

3. Such propositions relative to the uniformity of weights and measures as may be proper to be adopted in the United States.

The term *uniformity,* as applied to weights and measures, is susceptible of various constructions and modifications, some of which would restrict, while others would enlarge, the objects in contemplation by the resolution of the House.

Uniformity in weights and measures may have reference
1. To the weights and measures themselves.
2. To the objects of admeasurement and weight.
3. To *time,* or the duration of their establishment.
4. To *place,* or the extent of country over which, including the persons by whom, they are used.
5. To numbers, or the modes of numeration, multiplication, and division, of their parts and units.
6. To their *nomenclature,* or the denominations by which they are called.
7. To their connection with *coins* and *moneys of account.*

In reference to the weights and measures themselves, there may be

An uniformity of identity, or

An uniformity of proportion.

By an uniformity of *identity,* is meant a system founded on the principle of applying only one unit of weights to all weighable articles, and one unit of measures of capacity to all substances, thus measured, liquid or dry.

By an uniformity of *proportion,* is understood a system admitting more than one unit of weights, and more than one of measures of capacity; but in which all the weights and measures of capacity are in a uniform proportion with one another.

Our present existing weights and measures are, or orig-

inally were, founded upon the uniformity of proportion.
The new French metrology is founded on the uniformity
of identity.

And, in reference to each of these circumstances, and to
each in combination with all, or either of the others, uniformity may be more or less extensive, partial, or complete.

Measures and weights are the instruments used by man
for the comparison of quantities, and proportions of things.

In the order of human existence upon earth, the objects
which successively present themselves, are man—natural,
domestic, civil society, government, and law. The want,
at least, of measures of length, is founded in the physical
organization of individual man, and precedes the institution of society. Were there but one man upon earth, a
solitary savage, ranging the forests, and supporting his existence by a continual conflict with the wants of his nature
and the rigor of the elements, the necessities for which he
would be called to provide would be *food, raiment, shelter.*
To provide for the wants of food and raiment, the first occupation of his life would be the chase of those animals,
the flesh of which serves him for food, and the skins of
which are adaptable to his person for raiment. In adapting the raiment to his body, he would find at once, in his
own person, the want and the supply of a standard measure
of length, and of the proportions and subdivisions of that
standard.

But, to the continued existence of the human species,
two persons of different sexes are required. Their union
constitutes natural society, and their permanent cohabitation, by mutual consent, forms the origin of domestic society. Permanent cohabitation requires a common place

of abode, and leads to the construction of edifices where the associated parties, and their progeny, may abide. To the construction of a dwelling-place, superficial measure becomes essential, and the dimensions of the building still bear a natural proportion to those of its destined inhabitants. Vessels of capacity are soon found indispensable for the supply of water; and the range of excursion around the dwelling could scarcely fail to suggest the use of a measure of itinerary distance.

Measures of *length*, therefore, are the wants of individual man, independent of, and preceding, the existence of society. Measures of surface, of distance, and of capacity arise immediately from domestic society. They are wants proceeding rather from social, than from individual, existence. With regard to the first, *linear* measure, nature in creating the want, and in furnishing to man, within himself, the means of its supply, has established a system of numbers, and of proportions, between the man, the measure, and the objects measured. Linear measure requires only a change of direction to become a measure of circumference; but is not thereby, without calculation, a measure of surface. Itinerary measure, as it needs nothing more than the prolongation or repetition of linear measure, would seem at first view to be the same. Yet this is evidently not the progress of nature. As the want of it originates in a different stage of human existence, it will not naturally occur to man, to use the same measure, or the same scale of proportions and numbers, to clothe his body and to mark the distance of his walks. On the contrary, for the measurement of all objects which he can lift and handle, the fathom, the arm, the cubit, the hand's-breadth, the span, and the fingers, are the instruments proposed to him by nature; while the pace and the foot are those which she gives him for the measurement of itinerary distance. These natural standards are never, in any stage of society, lost to individual man. There are probably few persons living who do not occasionally use their own arms, hands, and fingers, to measure objects which they handle, and their own pace to measure a distance upon the ground.

Here then is a source of *diversity*, to the standards even of linear measure, flowing from the difference of relations between man and physical nature. It would be as inconvenient and unnatural to the organization of the human body to measure a bow and arrow, for instance, the first furniture of solitary man, by his foot or pace, as to measure the distance of a day's journey, or a morning's walk to the hunting-ground, by his arm or hand.

Measures of capacity are rendered necessary by the nature of fluids, which can be held together in definite quantities only by vessels of substance more compact than their own. They are also necessary for the admeasurement of those substances which nature produces in multitudes too great for numeration, and too minute for linear measure. Of this character are all the grains and seeds, which, from the time when man becomes a tiller of the ground, furnish the principal materials of his subsistence. But nature has not furnished him with the means of supplying this want in his own person. For this measure he is obliged to look abroad into the nature of things; and his first measure of capacity will most probably be found in the egg of a large bird, the shell of a cetaceous fish, or the horn of a beast. The want of a *common* standard not being yet felt, these measures will be of various dimensions; nor is it to be expected that the thought will ever occur to the man of nature, of establishing a proportion between his cubit and his cup, of graduating his pitcher by the size of his foot, or equalizing its parts by the number of his fingers.

Measures of length, once acquired, may be, and naturally are, applied to the admeasurement of objects of surface and solidity; and hence arise new diversities from the nature of things. The connection of linear measure with *numbers*, necessarily, and in the first instance, imports only the first arithmetical rule of numeration, or addition. The mensuration of surfaces, and of solids, requires the further aid of multiplication and division. Mere numbers, and mere linear measure, may be reckoned by addition alone; but their application to the surface can be computed only by multiplication. The elementary principle of decimal arithmetic is then supplied by nature to man within himself in the number of his fingers. Whatever standard of linear measure he may assume, in order to measure the surface or the solid, it will be natural to him to stop in the process of addition when he has counted the tale equal to that of his fingers. Then turning his line in the other direction, and stopping at the same term, he finds the square of his number a hundred: and, applying it again to the solid, he finds its cube a thousand.

But while decimal arithmetic thus, for the purposes of *computation*, shoots spontaneously from the nature of man and of things, it is not equally adapted to the numeration, the multiplication, or the division of material substances.

either in his own person, or in external nature. The proportions of the human body, and of its members, are in other than decimal numbers. The first unit of measures, for the use of the hand, is the *cubit,* or extent from the tip of the elbow to the end of the middle finger; the motives for choosing which, are, that it presents more definite terminations at both ends than any of the other superior limbs, and gives a measure easily handled and carried about the person. By doubling this measure is given the ell, or arm, including the hand, and half the width of the body, to the middle of the breast; and, by doubling that, the fathom, or extent from the extremity of one middle finger to that of the other, with expanded arms, an exact equivalent to the stature of man, or extension from the crown of the head to the sole of the foot. For subdivisions and smaller measures, the span is found equal to half the cubit, the palm to one-third of the span, and the finger to one-fourth of the palm. The cubit is thus, for the mensuration of matter, naturally divided into 24 equal parts, with subdivisions of which 2, 3, and 4, are the factors; while, for the mensuration of distance, the foot will be found at once equal to one-third of the pace, and one-sixth of the fathom.

Nor are the diversities of nature, in the organization of external matter, better suited to the exclusive use of decimal arithmetic. In the three modes of its extension, to which the same linear measure may be applied, length, breadth, and thickness, the proportions of surface and solidity are not the same with those of length: that which is decimal to the line, is centesimal to the surface, and millesimal to the cube. Geometrical progression forms the rule of numbers for the surface and the solid, and their adaptation to decimal numbers is among the profoundest mysteries of mathematical science, a mystery which had been impenetrable to Pythagoras, Archimedes, and Ptolemy; which remained unrevealed even to Copernicus, Galileo, and Kepler, and the discovery and exposition of which was reserved to immortalize the name of Napier. To the mensuration of the surface, and the solid, the number ten is of little more use than any other. The numbers of each of the two or three modes of extension must be multiplied together to yield the surface or the solid contents: and, unless the object to be measured is a perfect square or cube of equal dimensions at all its sides, decimal arithmetic is utterly incompetent to the purpose of their admeasurement.

Linear measure, to whatever modification of matter applied, extends in a straight line; but the modifications of matter, as produced by nature, are in forms innumerable, of which the defining outward line is almost invariably a curve. If decimal arithmetic is incompetent even to give the dimensions of those artificial forms, the square and the cube, still more incompetent is it to give the circumference, the area, and the contents, of the circle and the sphere.

There are three several modes by which the quantities of material substances may be estimated and compared: by number, by the space which they occupy, and by their apparent specific gravity. We have seen the origin and character of mensuration by space and number, and that, in the order of human existence, one is the result of a necessity incidental to individual man preceding the social union, and the other immediately springing from that union. The union of the sexes constitutes natural society: their permanent cohabitation is the foundation of domestic society, and leads to that of government, arising from the relations between the parents and the offspring which their union produces. The relations between husband and wife import domestic society, consent, and the sacred obligation of promises. Those between parent and child, import subordination and government; on the one side authority, on the other obedience. In the first years of infancy the authority of the parent is absolute; and has, therefore, in the laws of nature, been tempered by parental affection. As the child advances to mature age, the relations of power and subjection gradually subside, and, finally, are dissolved in that honor and reverence of the child for the parent which can terminate only with life. When the child goes forth into the world to make a settlement for himself, and found a new family, civil society commences; government is instituted—the tillage of the ground, the discovery and use of metals, exchanges, traffic by barter, a *common* standard of measures, and mensuration by *weight*, or apparent specific gravity, all arise from the multiplying relations between man and man, now superadded to those between man and things.

The difference between the specific gravities of different substances is so great, that it could not, for any length of time, escape observation; but nature has not furnished man, within himself, with any standard for this mode of estimating equivalents. Specific gravity, as an object of mensuration, is in its nature *proportional*. It is not, like measures of length and capacity, a comparison between

different definite portions of space, but a comparison between different properties of matter. It is not the simple
relation between the extension of one substance and the
extension of another, but the complicated relation of extension and gravitation in one substance to the extension
and gravitation of another. This distinction is of great
and insuperable influence upon the principle of *uniformity*, as applicable to a system of weights and measures.
Extension and *gravitation* neither have nor admit of one
common standard. *Diversity* is the law of their nature,
and the only *uniformity* which human ingenuity can
establish between them is, an uniformity of proportion,
and not an uniformity of identity.

The necessity for the use of *weights* is not in the
organization of individual man. It is not essential even
to the condition or the comforts of domestic society. It
presupposes the discovery of the properties of the balance;
and originates in the exchanges of traffic, after the institution of civil society. It results from the experience
that the comparison of the articles of exchange, which
serve for the subsistence or the enjoyment of life, by their
relative extension, is not sufficient as a criterion of their
value. The first use of the balance, and of weights, implies
two substances, each of which is the test and the standard
of the other. It is natural that these substances should
be the articles the most essential to subsistence. They
will be borrowed from the harvest and the vintage: they
will be corn and wine. The discovery of the metals, and
their extraction from the bowels of the earth, must, in the
annals of human nature, be subsequent, but proximate,
to the first use of weights; and, when discovered, the only

mode of ascertaining their definite quantities will be soon perceived to be their weight. That they should, themselves, immediately become the common standards of exchanges, or otherwise of value and of weights, is perfectly in the order of nature; but their proportions to one another, or to the other objects by which they are to be estimated, will not be the same as standards of weight and as standards of value. Gold, silver, copper, and iron, when balanced each by the other in weight, will present masses very different from each other in value. They give rise to another complication, and another **diversity, of weights and** measures, equally **inaccessible to the uniformity** of identity and to the computations of decimal arithmetic.

Of the metals, that which, by **the adaptation of its** properties to the various uses of society, and to the purposes of traffic, by the quantities in which nature has disclosed it to the possession of man, intermediate between her profuse bounties of the coarser and her parsimonious dispensation of the finer metals, holds a middle station between them, wins its way **as** the common, **and at last as** the only, standard of value. **It** becomes the universal **medium of** exchanges. Its quantities, ascertained by weight, become themselves the standards of weights. Civil government is called in as the guardian and voucher of its purity. **The** civil authority stamps its image, to authenticate **its** weight and alloy; and silver becomes at **once a** weight, money, and coin.

With civil society, too, originates the necessity for common and uniform standards of measures. Of the different measures of extension necessary for individual man, and **for** domestic society, although the want will be common

to all and frequently recurring, yet the standards will not
be uniform, either with reference to time or to persons.
The standard of linear measure for each individual being
in himself, those of no two individuals will be the same.
At different times the same individual will use different
measures, according to the several purposes for which they
will be wanted. In domestic society, the measures adaptable to the persons of the husband, of the wife, and of the
children, are not the same ; nor will the idea of reducing
them all to one common standard press itself upon their
wants, until the multiplication of families gives rise to
the intercourse, exchanges, and government of civil society. Common standards will then be assumed from the
person of some distinguished individual ; but accidental
circumstances, rather than any law of nature, will determine whether identity or proportion will be the character
of their uniformity. If, pursuing the first and original
dictate of nature, the cubit should be assumed as the
standard of linear measure for the use of the hand, and
the pace for the measure of motion, or linear measure
upon earth, there will be two units of long measure—one
for the measure of matter, and another for the measure
of motion. Nor will they be reducible to one ; because
neither the cubit nor the pace is an aliquot part or a multiple of the other. But, should the discovery have been
made that the *foot* is at once an aliquot part of the pace
for the mensuration of motion, and of the ell and fathom
for the mensuration of matter, the foot will be made the
common standard measure for both ; and, thenceforth,
there will be only one standard unit of long measure, and
its uniformity will be that of identity.

Thus, in tracing the theoretic history of weights and measures to their original elements in the nature and the necessities of man, we have found linear measure with individual existence, superficial, capacious, itinerary measure, and decimal arithmetic, with domestic society; weights and common standards, with civil society; money, coins, and all the elements of uniform metrology, with civil government and law; arising in successive and parallel progression together.

When weights and measures present themselves to the contemplation of the legislator, and call for the interposition of law, the first and most prominent idea which occurs to him is that of *uniformity:* his first object is to embody them into a system, and his first wish, to reduce them to one universal common standard. His purposes are uniformity, permanency, universality; one standard to be the same for all persons and all purposes, and to continue the same forever. These purposes, however, require powers which no legislator has hitherto been found to possess. The power of the legislator is limited by the extent of his territories, and the numbers of his people. His principle of universality, therefore, cannot be made, by the mere agency of his power, to extend beyond the inhabitants of his own possessions. The power of the legislator is limited over time. He is liable to change his own purposes. He is not infallible: he is liable to mistake the means of effecting his own objects. He is not immortal: his successor succedes to his power, with different views, different opinions, and perhaps different principles. The legislator has no power over the properties of matter. He cannot give a new constitution to nature. He cannot repeal her law of universal mutability. He cannot square the circle. He cannot reduce extension and gravity to one common measure. He cannot divide or multiply the parts of the surface, the cube, or the sphere, by the uniform and exclusive number ten. The power of the legislator is limited over the will and actions of his subjects. His conflict with them is desperate, when he counteracts their settled habits, their established usages; their domestic and individual economy, their ignorance, their prejudices, and their wants: all which is unavoidable in the attempt radically to change, or to originate, a totally new system of weights and measures.

In the origin of the different measures and weights, at different stages of man's individual and social existence; in the different modes by which nature has bounded the extension of matter; in the incommensurable properties of the straight and the curve line; in the different properties of matter, number, extension, and gravity, of which measures and weights are the tests, nature has planted sources of diversity, which the legislator would in vain overlook, which he would in vain attempt to control. To these sources of diversity in the nature of things, must be added all those arising from the nature and history of man. In the first use of weights and measures, neither universality nor permanency are essential to the uniformity of the standards. Every individual may have standards of his own, and may change them as convenience or humor may dictate. Even in civil society, it is not *necessary*, to the purposes of traffic, that the standards of the buyer and seller should be the same. It suffices, if the proportions between the standards of both parties are mutually understood. In the progress of society, the use of weights and measures having preceded legislation, if the families, descended from one, should, as they naturally may, have the same standards, other families will have others. Until regulated by law, their diversities will be numberless, their changes continual.

These diversities are still further multiplied by the abuses incident to the poverty, imperfections, and deceptions of human language. So arbitrary and so irrational is the dominion of usage over the speech of man, that, instead of appropriating a specific name to every distinct thing, he is impelled, by an irresistible propensity, sometimes to give different names to the same thing, but far more frequently to give the same name to different things. Weights and measures are, in their nature, relative. When man first borrows from his own person a standard measure of length, his first error is to give to the measure the name of the limb from which it is assumed. He calls the *measure* a cubit, a span, a hand, a finger, or a foot, improperly applying to it the name of those respective parts of his body. When he has discovered the properties of the balance, he either confounds with it the name of the weight, which he puts in it to balance the article which he would measure, or he gives to the definite mass, which he assumes for his standard, the indefinite and general name of *the weight.* Such was the original meaning of the weight which we call a *pound.* But, as different families assume different masses of gravity for their unit of weight, the pound of one bears the same name, and is a very different thing from the pound of another. When nations fall into the use of different weights or measures for the estimation of different objects, they commit the still grosser mistake of calling several different weights or measures by the same name. And, when governments degrade themselves by debasing their coins, as unfortunately all governments have done, they add the crime of fraud to that of injustice, by retaining the name of things which they have destroyed or changed. Even things which Nature has discriminated so clearly that they cannot be mistaken, the antipathy of mankind to new words will misrepresent and confound. It suffers not even numbers to retain their essentially definite character. It calls sixteen a dozen. It makes a hundred and twelve a hundred, and twenty-eight, twenty-five. Of all the tangles of confusion to be unravelled by the regulation of weights and measures, these abuses of language in their nomenclature are perhaps the most inextricable. So that when law comes to establish its principles of permanency, uniformity, and universality, it has to contend not only with the diversities arising from the nature of things and of man; but with those, infinitely more numerous, which proceed from existing usages, and delusive language; with the partial standards, and misapplied names, which have crept in with the lapse of time, beginning with individuals or families, and spreading more or less extensively to villages and communities.

In this conflict between the dominion of usage and of law, the last and greatest dangers to the principle of uniformity proceed from the laws themselves. The legislator having no distinct idea of the uniformity of which the subject is susceptible, not considering how far it should be extended, or where it finds its boundary in the nature

of things and of man, enacts laws inadequate to their purpose, inconsistent with one another; sometimes stubbornly
resisting, at others weakly yielding to inveterate uses or
abuses; and finishes by increasing the diversities which it
was his intention to abolish, and by loading his statute-
book only with the impotence of authority, and the uniformity of confusion.

This inquiry into the theory of weights and measures,
as resulting from the natural history of man, was deemed
necessary as preliminary to that statement of the proceedings of foreign countries for establishing uniformity in
weights and measures, called for by the resolution of the
House.

It presents to view certain principles believed to be
essential to the subject, upon which the historical statement required will shed continual illustration, and which
it will be advisable to bear in mind, when the propositions
supposed to be proper for the adoption of the United
States are to be considered.

In this review, civil society has been considered as originating in a single family. It can never originate in any
other manner. But government, and national communities, may originate either by the multiplication of families
from one, or in compact, by the voluntary association of
many families, or in force, by conquest. In the nations
formed by the reunion of many families, each family will
have its standard measures and weights already settled, and
common standards for the whole can be established only
by the means of *law*. It is a consideration from which
many important consequences result, that the proper province of law, in relation to weights and measures, is, not to

create, but to regulate. It finds them already existing, with diversities innumerable, arising not only from all the causes which have been enumerated, but from all the frauds to which these diversities give continual occasion and temptation.

There are two nations of antiquity from whom almost all the civil, political, and religious institutions of modern Europe, and of her descendants in this hemisphere, are derived—the Hebrews, and the Greeks. They both, at certain periods, not very distant from each other, issued from Egypt; and both nearly at the time of the first invention of alphabetical writing. The earliest existing records of history are of them, and in their respective languages. They exhibit examples of national communities and governments originating in two of the different modes noticed in the preceding remarks. The Hebrews sprung from a single family, of which Abraham and Sarah were the first founders. The Greeks were a confederated nation, formed by the voluntary association of many families. To their historical records, therefore, we must appeal for the actual origin of our own existing weights and measures; and, beginning with the most ancient of them, the Hebrews, it is presumed, that the Scriptures may be cited in the character of historical documents. We there find, that all the human inhabitants of this globe sprung from one created pair; that the necessity of raiment adapted to the organization of their bodies, and of the tillage of the ground for their subsistence, arose by their fall from innocence; that their eldest son was a tiller of the ground, and built a city, and their second son a keeper of sheep; that, at no distant period from the creation, instruments of brass and iron

were invented. Of the origin of weights and measures no
direct mention is made; but the Hebrew historian, Jose-
phus, asserts, that they were invented by Cain, the tiller
of the ground, and the first builder of a city. As the duration of human life was tenfold longer before the flood
than in later ages, the multiplication of the species was
proportionally rapid; and the inventions and discoveries
of many ages were included within the life of every individual. In the early stages of man's existence upon earth,
direct revelations from the Creator were also frequent, and
imparted knowledge unattainable but in a series of centuries to the merely natural energies of the human mind.
The division of numbers by decimal arithmetic, and the
use of the *cubit* as a standard measure of length, are distinctly proved to have been established before the general
deluge. The division of time into days, months, and
years, was settled. The ages of the patriarchs are noted
in units, tens, and hundreds of years; and Noah, we are
told, built, by divine instruction, his ark three hundred
cubits long, fifty cubits broad, and thirty cubits in height.

After the general deluge, the dispersion of the human
species, and the confusion of languages which ensued,
must have destroyed whatever uniformity of weights and
measures might have existed, while the whole earth was
of one language and of one speech. After noticing this
great and miraculous event, the historical part of the Bible
is chiefly confined to the family of Abraham, originally a
Chaldean, said to have been very rich in cattle, in silver,
and in gold. In his time, we find mention made of *meas-*
ures of meal. Abimelech gives him a thousand pieces of
silver. He, himself, gives to Hagar a *bottle* of water and

buys of Ephron, the Hittite, the field of Machpelah, for which he pays him, by *weight,* **four** hundred shekels of silver, current money with the **merchant.** At this period, therefore, we find established measures of length of land, and of capacity, liquid and dry; weights, **coined** money, and decimal arithmetic. The elements for **a** system of metrology are complete; but the only uniformity **observable** in them is the identity of weights and coin, and **the** decimal numbers.

In the *law* given from Sinai—the law, not of a human legislator, but of God—there are two precepts respecting weights and measures. The first [Leviticus xix. 35, 36], "Ye shall do no unrighteousness in judgment, in mete-yard (measure of length), in weight, or in measure (of capacity). Just balances, just weights, a just ephah, and **a just hin shall** ye have." **The second** [Deuteronomy **xxv. 13, 14, 15],** "Thou shalt not have in thy bag *divers* weights, a great and a small. Thou shalt **not** have in thine house divers measures, a great and **a small.** But thou shalt have a perfect and just weight, a perfect and just measure shalt thou have." The weights and measures are prescribed as already existing and known, and were all probably the same as those of the Egyptians. The first of these injunctions is addressed in the plural to the whole nation, and the second in the singular to every individual. The first has reference to the standards, which were to be kept in the ark of the covenant, or the sanctuary; and the second to **the** copies of them, kept by every **family** for their own use. The first, therefore, only comman**ds** that the standards should be *just:* and that, in all transactions, for which weights and measures might **be** used, the principle of righteousness should be observed. The second requires, that the copies of the standards used by individuals, should be uniform, not divers; and not only just, but perfect, with reference to the standards.

The long measures were, the *cubit*, with its subdivisions of two *spans*, six palms or hand-breadths, and twenty-four digits or fingers. It had no division in decimal parts, and was not employed for itinerary measure: that was reckoned by paces, Sabbath-day's journeys, and day's journeys. The measures of capacity were, the ephah for the dry, and the *hin* for liquid measure; the primitive standard from nature of which was an egg-shell; six of these, constituted the *log*, a measure little less than our *pint*. The largest measure of capacity, the *homer*, was common both to liquid and dry substances; and its contents nearly corresponded with our wine hogshead, and with the Winchester quarter. The intermediate measures were different, and differently subdivided. They combined the decimal and duodecimal divisions: the latter of which may, perhaps, have arisen from the accidental number of the tribes of Israel. Thus, in liquids, the bath was a tenth part of the homer, the hin a sixth part of the bath, and the log a twelfth part of the hin; while, for dry measure, the ephah was a tenth part of the homer, the seah a third, and the omer a tenth part of the ephah; and the cab a sixth part of the seah. The weights and coins were, the shekel, of twenty gerahs; the maneh, which for weight was of sixty and in money of fifty shekels; and the kinchar, or talent, of three thousand shekels in both. The ephah had also been formed by the process of cubing an Egyptian measure of length called the *ardob*. The original weight of the shekel was the same as one-half of our avoirdupois ounce; the most ancient of weights traceable in human history.

And thus the earliest and most venerable of historical records extant, in perfect coincidence with speculative theory, prove, that the natural standards of weights and measures are not the same; that even the natural standards of cloth and of long measure are two, both derived from the stature and proportions of man, but one from his hand and arm, and the other from his leg and foot; that the natural standards of measures of capacity and of weights are different from those of linear measure, and different from each other, the essential character of the weight being compact solidity, and that of the vessel bounded vacuity; that the natural standards of weights are two, one of which is the same with metallic money; and that decimal arithmetic, as founded in nature, is peculiarly applicable to the standard *units* of weights and measures, but not to their subdivisions or fractional parts, nor to the objects of admeasurement and weight.

With all these diversities, the only commands of the law for observing uniformity were, that the weights and the measures should be just, perfect, and not divers, a *great* and a *small*. But this last prohibition was merely an ordinance against fraud. It was a precept to the individual, and not to the nation. It forbade the iniquitous practice of using a large weight or measure for buying, and a small one for selling the same article; and to remove the opportunity for temptation, it enjoined upon the individual not to have divers weights and measures, great and small, of the same denomination, in his bag when at market, or in his house when at home. But it

was never understood to forbid that there should be measures of different dimensions bearing the same name : and
it appears, from the sacred history, that there actually were
three different measures called a cubit, of about the relative proportion of 17, 21, and 35, of our inches, to each
other. They were distinguished by the several denominations of the cubit of a man, the cubit of the king, and the
cubit of the sanctuary.

In the vision of the prophet Ezekiel, during the Baby-
lonian captivity—that vision which, under the resurrection
of dead bones, shadowed forth the restoration and union
of the houses of Ephraim and of Judah—with the re-
proaches of former violence and spoil, injustice and exactions, are mingled the exhortations of future righteousness, particularly with reference to weights and measures:
and there is a special command that the measures of capacity, liquid and dry, should be the same.

"Thus saith the Lord God: let it suffice you, O princes
of Israel, remove violence and spoil, and execute judgment
and justice, take away your exactions from my people,
saith the Lord God. Ye shall have just balances, and a
just ephah, and a just bath. The ephah and the bath
shall be one measure, that the bath may contain the tenth
part of an homer, and the ephah the tenth part of an
homer; the measure thereof shall be after the homer.
And the shekel shall be twenty gerahs: twenty shekels,
five and twenty shekels, fifteen shekels, shall be your
maneh. This is the oblation that ye shall offer: the sixth
part of an ephah of an homer of wheat; and ye shall give
the sixth part of an ephah of an homer of barley. Concerning the ordinance of oil, ye shall offer the tenth part

of a bath out of the cor, which is an homer of ten baths; for ten baths are an homer."

Here we see combined the uniformity of identity and the uniformity of proportion. The homer was a dry, and the cor a liquid, measure of capacity: they were of the same contents; the ephah and the bath were their corresponding tenth parts, also of the same capacity. But the oblation of wheat and barley was to be a *sixth* part of the ephah, and the oblation of oil a tenth part of the bath. The oblations were uniform, but the measures were proportional; and that proportion was compounded of the different weight and value of the respective articles.

In the same vision of Ezekiel, the directions are given for the building of the new temple after the restoration of the captivity; and all the dimensions of the temple are prescribed by a measuring reed of six cubits long by the cubit and an hand-breadth. "And these (says he) are the measures of the altar after the cubits: *the cubit is a cubit and an hand-breadth.*" (Ch. xliii. 13.)

The book of Job is a story of a man supposed not to have been descended from Abraham, and certainly not belonging to any of the tribes of Israel. It has reference to other manners, other customs, opinions, and laws, than those of the Hebrews. But it bears evidence of the primitive custom of paying silver by weight, while gold and jewels were valued by tale; and of that system of proportional uniformity which combines gravity and extension for the mensuration of fluids. Speaking of wisdom, it says (ch. xxviii. 15, 17), "It cannot be gotten for gold, neither *shall silver be weighed* for the price thereof. The gold and the crystal cannot equal it; and the exchange of it shall not be **for** jewels of fine gold." And afterward, in the same chapter, that "God maketh the weight for the winds, *and weigheth the waters by measure.*"

The cubit was also a primitive measure of length among the Greeks; but, at the institution of the Olympic games by Hercules, his foot is said to have been substituted as the unit of measure for the foot-race. Six hundred **of** these feet constituted the stadium, or length of the course or stand, which thenceforth became the standard itinerary **measure of the** nation. It was afterward by the Romans **combined with** the pace, **a thousand** of which constituted **the mile. The** foot and the mile, or thousand **paces, are our** standard measures of length **at this day.**

The foot has over the cubit the advantage **of being a** common aliquot part both of the pace and the fathom. It is also definite at both extremities, and affords the natural means of reducing the two standard measures of length to one. Its adoption was therefore a great and important advance toward uniformity; and **this may ac-**count for the universal abandonment, by all **the** modern nations of Europe, of that primitive antediluvian **standard** measure, the *cubit.*

Of the Greek weights and measures of capacity the origin is not distinctly known; but that whatever uniformity ever existed in the system was an uniformity of proportion, and not of identity, is certain. They had weights corresponding to our avoirdupois and troy pounds, and measures answering to our wine and ale gallons; **not** indeed in the same proportions, but in the proportions to **each** other of the weight of wine and oil.

It has been observed that the process of weighing implies two substances, each of which is the standard **and** test of the other; **that, in the** order of human existence, the use **of weights precedes** the weighing of metals, but that, when the metals and their uses to **the** purposes of life are discovered, their value can at first be estimated **only by weight, whereby they soon become** standards both **of weight and of value for all other things. This theory is** confirmed by the history of the **Greek, no less than by that of the** Hebrew, weights and measures. The term talent, in its primitive meaning in the Greek language, signified **a** *balance;* and it was at once the largest weight **and the** highest denomination of money among the Greeks. **Its** subdivisions, the *mina* and the *drachma,* were at once weights and money; and the ***drachma*** was the unit of all the silver coins. **But the money which was** a weight, **though substituted for many purposes, instead of the more ancient weight by which it had itself been tried, never ex-**cluded it from **use. It had** not the **fortune of the** *foot,* **to** banish from **the use** of mankind its predecessor. **They** had the weight for money, and the weight for **measure.**

As there are thus in nature two standards of **weight,** there are also two of **measures** of capacity. From the names **of the** Greek measures of capacity, they were originally assumed from cockle **and other** shells of fish. But as **these give no** scales of proportion for subdivisions, when **reduced to** a system, their capacity was determined by the **two** modifications of matter, extension and weight. Like **the** Hebrews, they had measures for liquid **and dry** sub**stances, which** were the same, but with different multiples and subdivisions. Their measures of **wine** and oil were determined by the *weight* of their contents; their measures of water and of grain, by vessels of capacity cubed from measures of length.

The weights and measures of the Romans were all derived from those of the Greeks. The identity of one of their standard units of weight, with money and coin, was the same. *Aes*, brass, was their original money: and as its payment was by weight, the term pound, *libra*, was the balance; and *money* was the weight of brass in the balance. The general term soon came to be applied to a definite weight: and when afterward silver came to be coined, the *sestertius*, which signified two and a half, and the *denarius*, or piece of *ten*, meant the pieces of silver of value equal respectively to two and a half and to ten of the original brass weights of the balance. The sestertius was the unit of money, and the denarius of silver coins.

The Romans had also two pound-weights; which were termed the metrical and the scale pound. "The scale pound," says Galen, "determines the *weight* of bodies; the metrical pound, the contents or quantity of *space* which they fill."

Their measures of capacity for wet or dry substances were in like manner, in part, the same, but with different multiples and subdivisions. Like them they were formed of the two different processes of cubing the foot, and of testing wine and oil by weight. The *amphora*, or largest measure of liquids, weighed eighty pounds of water, and was formed by cubing, or, as they called it, squaring their foot measure: it was for that reason called a *quadrantal*. But their *congius*, or unit of liquid measure, was any vessel containing ten metrical *pounds weight* of wine. The Silian law, enacted nearly three centuries before the Christian era, expressly declares that the quadrantal contains eighty *pounds* of wine, the *congius* ten pounds; that the *sextarius* contains the sixth *part* of a congius, and is a measure both for liquid and dry substances; that forty-eight sextarii make a quadrantal of wine, and sixteen *libræ* a modius. The money pound, or *pondo,* and the metrical pound, or *libra,* were in the proportion to each other, of 84 to 100, nearly the same as that between our troy and avoirdupois weights. [Arbuthnot on Coins, Weights, and Measures, p. 23.] There is a standard congius of the age of Vespasian still extant at Rome; and the inscription upon it marks, that it contains ten pounds of wine.

Among the nations of modern Europe there are two, who, by their genius, their learning, their industry, and their ardent and successful cultivation of the arts and sciences, are scarcely less distinguished than the Hebrews from whom they have received most of their religious, or the Greeks from whom they have derived many of their civil and political institutions. From these two nations the inhabitants of these United States are chiefly descended; and from one of them we have all our existing weights and measures. Both of them, for a series of ages, have been engaged in the pursuit of an uniform system of weights and measures. To this the wishes of their philanthropists, the hopes of their patriots, the researches of their philosophers, and the energy of their legislators, have been aiming with efforts so stupendous and with perseverance so untiring, that, to any person who shall examine them, it may well be a subject of astonishment to find that they are both yet entangled in the pursuit at this hour, and that it may be doubted whether all their latest

and greatest exertions have not hitherto tended to increase
diversity instead of producing uniformity.

It was observed, at the introduction of these remarks,
that one of the primary elements of uniformity, as applied
to a system of weights and measures, has reference to the
persons by whom they are used; and it has since been
noticed, that the power of the legislator is restricted to
the inhabitants of his own dominions. Now, the perfection of uniformity with respect to the persons to whose
use a system of metrology is adapted, consists in its embracing, at least in its aptitude, the whole human race.
In the abstract, that system which would be most useful
for one nation, would be the best for all. But this uniformity cannot be obtained by legislation. It must be
imposed by conquest, or adopted by consent. When therefore two populous and commercial nations are at the same
time forming and maturing a system of weights and measures on the principle of uniformity, unless the system
proves to be the same, the result as respects all their relations with each other must be, not uniformity, but new
and increased diversity. This consideration is of momen-
tous importance to the people of this Union. Since the
establishment of our national independence, we have par-
taken of that ardent spirit of reform, and that impatient
longing for uniformity, which have so signally animated
the two nations from whom we descended. The Congress
of the United States have been as earnestly employed in
the search of an uniform system of weights and measures
as the British Parliament. Have either of them considered,
how that very principle of uniformity would be affected
by any, the slightest change, sanctioned by either, in the

existing system, now common to both? If uniformity be
their object, is it not necessary to contemplate it in all its
aspects? And while squaring the circle to draw a straight
line from a curve, and fixing mutability to find a standard
pendulum, is it not worth their while to inquire, whether
an imperceptible improvement in the uniformity of things
would not be dearly purchased by the loss of millions in
the uniformity of persons?

It is presumed that the intentions of the House, in requiring a statement of the proceedings in foreign countries
for establishing uniformity in weights and measures will
be fulfilled by confining this part of the inquiry to the
proceedings of the two nations above mentioned. It
appears that a reformation of the weights and measures
of Spain is among the objects now under the consideration of the Cortes of that kingdom: and, as weights and
measures are the necessary and universal instruments of
commerce, no change can be effected in the system of any
one nation without sensibly affecting, though in very different degrees, all those with whom they entertain any
relations of trade. But the results of this inquiry, newly
instituted in Spain, have not yet been made known.
France and Great Britain are the only nations of modern
Europe who have taken much interest in the organization
of a new system, or attempted a reform for the avowed
purpose of uniformity. The proceedings in those two
countries have been numerous, elaborate, persevering,
and, in France especially, comprehensive, profound, and
systematic. In both, the phenomenon is still exhibited,
that, after many centuries of study, of invention of laws,
and of penalties, almost every village in the country is in

the habitual use of different weights and measures; which diversity is infinitely multiplied, by the fact, that, in each country, although the quantities of the weights and measures are thus different, their denominations are few in number, and the same names, as foot, pound, ounce, bushel, pint, etc., are applied in different places, and often in the same place, to quantities altogether diverse.

During the conquering period of the French Revolution, the new system of French weights and measures was introduced into those countries which were united to the empire. Since the severance of those countries from France, it has been discarded, excepting in the kingdom of the Netherlands, where, by two ordinances of the king, it has been confirmed with certain exceptions and modifications, particularly with regard to the coins.

In England, from the earliest records of parliamentary history, the statute-books are filled with ineffectual attempts of the legislature to establish uniformity. Of the origin of their weights and measures, the historical traces are faint and indistinct; but they have had, from time immemorial, the pound, ounce, foot, inch, and mile, derived from the Romans, and through them from the Greeks, and the *yard*, or *girth*, a measure of Saxon origin, derived, like those of the Hebrews and the Greeks, from the human body, but, as a natural standard, different from theirs, being taken not from the length or members, but from the circumference of the body. The yard of the Saxons evidently belongs to a primitive system of measures different from that of the Greeks, of which the foot, and from that of the Hebrews, Egyptians, and antediluvians, of which the cubit was the standard. It affords, therefore, another demonstration, how invariably nature first points to the human body, and its proportions, for the original standards of linear measure. But the *yard* being, for all purposes of use, a measure corresponding with the *ulna*, or ell, of the Roman system, became, when superadded to it, a source of diversity, and an obstacle to uniformity in the system. The yard, therefore, very soon after the Roman conquest, is said to have lost its original character of girth; to have been adjusted as a standard by the arm of King Henry the First; and to have been found or made a multiple of the foot, thereby adapting it to the remainder of the system: and this may perhaps be the cause of the difference of the present English foot from that of the Romans, by whom, as a measure, it was introduced. The ell measure has, however, in England, retained its place as a standard for measuring cloth; but, in the ancient statutes, which for centuries after the conquest were enacted in the degenerate Latin of the age, the term *ulna*, or ell, is always used to designate the yard. Historical traditions allege that, a full century before the Conquest, a law of Edgar prescribed that there should be the same weights and the same measures throughout the realm, but that it was never observed. The system which had been introduced by the Romans, however uniform in its origin, must have undergone various changes in the different governments of the Saxon heptarchy. When those kingdoms were united in one, it was natural that laws of uniformity should be prescribed by the prince, and as natural that usages of diversity should be persisted in by the people. Canute the Dane, William the Conqueror, and Richard the First, princes among those of most extensive and commanding authority, are said to have made laws of the like import and the same inefficacy. The Norman Conquest made no changes in any of the established weights and measures. The very words of a law of William the Conqueror are cited by modern writers on the English weights and measures; their import is: "We ordain and command that the weights and measures, throughout the realm, be as our worthy predecessors have established." [Wilkins, Legg, Saxon, Folkes, cited by Clark, p. 150.]

One of the principal objects of the Great Charter was the establishment of uniformity of weights and measures; but it was a uniformity of *existing* weights and measures, and a uniformity not of identity, but of proportion. The words of the 25th chapter of the Great Charter of the year 1225 (9 Henry III.) are, in the English translation of the statutes, "One measure of wine shall be through our realm, and one measure of ale, and one measure of corn, that is to say, the *quarter* of London: and one breadth of dyed cloth, that is to say, two yards (ulne) within the lists: and it shall be of weights as it is of measures." The London quarter, therefore, and the *yard*, or *ulna*, were existing, known, established measures; and the one measure of corn was the London quarter. The one measure of ale was a gallon, of the same contents for liquid measure as the half-peck was for dry. But the one measure of wine was a gallon, not of the same cubical contents as the half-peck and ale gallon, but which, when filled with wine, was of the same weight as the half-peck, or corn gallon, when filled with wheat. And the expressions, "it shall be of weights as it is of measures," mean that there shall be the same proportion between the money weight and the merchant's weight, as between the wine measure and the corn measure.

The Great Charter, which now appears as the first legislative act in the English statutes at large, is not the Magna Charta extorted by the barons from John at Runnimede, but a repetition of it by Henry the Third in the year 1225, as confirmed by his son, Edward the First, in the year 1300. It is properly an act of this last date, though inserted in the book as of 9 Henry III., or 1225.

In several of the subsequent confirmations of this charter, which for successive ages attest at once how apt it was to be forgotten by power, and how present it always was to the memory of the people, the real meaning of this 25th chapter appears to have been misunderstood. It has been supposed to have prescribed the uniformity of identity, and not the uniformity of proportion; that, by enjoining one measure of wine, and one measure of ale, and one measure of corn, its intention was that all these measures should be the same; that there should be only one unit of measure of capacity for liquid and dry substances, and one unit of weights.

But this neither was, nor could be, the meaning of the statute. Had it been the intention of the legislator, he would have said, there shall be one and the same measure for wine, corn, and ale; and the reference to the London quarter could not have been made, for neither wine nor ale were ever measured by the quarter; and, instead of saying "it shall be of weights as it is of measures," it would have said, there shall be but one set of weights for whatever is to be weighed.

The object of the whole statute was, not to innovate, but to fix existing rights and usages, and to guard against fraud and oppression. It says that the measure of corn shall be the London quarter; that cloth shall be two yards within the lists. But it neither defines the contents of the quarter, nor the length of the yard: it refers to both as fixed and settled quantities. To **have prescribed that** there should be **but** one unit of weights and one measure of wine, ale, and corn, would have been a great and violent innovation upon all the existing habits and usages of the people. The chapter is not intended for a *general* regulation of weights and measures. It refers specifically and exclusively to the measure of three articles, wine, **ale,** corn; and to the width of cloths. Its intention was **to** provide that the measure of corn, **of** ale, and of wine, should *not* be the same; that is, that the wine measure should not be used for ale and corn, nor the ale measure for wine.

That such was and must have been the meaning of the statute, is further proved by the statute **of 1266 (51** Henry III.), **and** by the treatise upon weights **and meas-**ures, published **in** the statute-books as of the 31 Edward I., or 1304; the first, an act of the same Henry the Third whose Great Charter is that inserted among the laws, and the **second** an act of the same Edward **the First** whose confirmation of the Great Charter is the existing statute.

The act of 51 Henry III. (1266), is called the assize of bread and of ale. It purports to be an exemplification, given at the request of the bakers of the town of Coventry, of certain ordinances, of the assize of bread, and ale, and of the making of money and measures, made in the times of the king's progenitors, sometime kings of England. It presents an established scale, then of ancient standing, between the prices of wheat and of bread, providing that when the *quarter* of wheat is sold at twelve pence, the farthing loaf of the best white bread shall weigh six pounds sixteen shillings. It then graduates the weight of bread according to the price of wheat, and for every six pence added to the quarter of wheat, reduces, though not in exact proportions, the weight of the farthing loaf, till, when the wheat is at twenty shillings a quarter, it directs the weight of the loaf to be six shillings and three pence. It regulates, in like manner, the price of the gallon of ale, by the price of wheat, barley, and oats; and, finally, declares that, "by the consent of *the whole realm* of England, the *measure* of the king was made; that is to say: that an English *penny*, called a sterling round, and without any clipping, shall weigh thirty-two wheat corns in the midst of the ear, and twenty-pence do make an ounce, and twelve ounces one pound, and eight pounds do make a gallon of *wine*, and eight gallons of *wine* do make a London bushel, which is the eighth part of a quarter."

Henry the Third was the eighth king of the Norman race: and this statute was passed exactly two hundred years after the Conquest. It is merely an exemplification, word for word, embracing several ordinances of his progenitors, kings of England; and it unfolds a system of uniformity for weights, coins, and measures of capacity, very ingeniously imagined, and skilfully combined.

It shows, first, that the money weight was identical with the silver coins: and it establishes an uniformity of proportion between the money weight and the merchant's weight, exactly corresponding to that between the measure of wine and the measure of grain.

It makes wheat and silver money, the two weights of the balance, the natural tests and standards of each other; that is, it makes wheat the standard for the weight of silver money, and silver money the standard for the weight of wheat.

It combines an uniformity of proportion between the weight and the measure of wheat and of wine; so that the measure of wheat should at the same time be a certain weight of wheat, and the measure of wine at the same time a certain weight of wine, so that the article whether bought and sold by weight or by measure, the result was the same. To this, with regard to wheat, it gave the further advantage of an abridged process for buying or selling it by the number of its kernels. Under this system, wheat was bought and sold by a combination of every property of its nature, with reference to quantity; that is, by number, weight, and measure. The statute also fixed its proportional weight and value with reference to the weight and value of the silver coin for which it was to be exchanged in trade. If, as the most eminent of the modern economists maintain, the value of everything in trade is regulated by the proportional value of money and of wheat, then the system of weights and measures, contained in this statute, is not only accounted for as originating in the nature of things, but it may be doubted whether any other system be reconcileable to nature. It was with reference to this system, that, in the introduction to this report, it was observed, that our own weights and measures were originally founded upon an uniformity of proportion, and not upon an uniformity of identity. In the system which allows only one unit of weights and one unit of measures of capacity, all the advantages of the uniformity of proportion are lost. The litre of the French system is a weight for nothing but distilled water, at a given temperature.

But with this statute of 1266, and with the admirable system of proportional uniformity in weights and measures, of which it gives the elements, it has fared still worse than with the twenty-fifth chapter of Magna Charta. The most valuable and important feature of uniformity in the system, the identity of the mummulary weight and of the standard silver coin, that feature which is believed to be of more influence upon the happiness and upon the morals of nations, than any other principle of uniformity of which weights and measures are susceptible, was first defaced by Edward the First himself. It was utterly annihilated by his successors. The consequence of which has been, that the object and scope of the statute of 1266 have been misunderstood by subsequent parliaments; that laws have been enacted professedly in conformity to this statute, but entirely subversive of it; and that anomalies have crept into the weights and measures of England, and of this Union, which it appears to be impossible to trace to any other source.

The only notice which most of the modern writers upon English weights and measures have taken of this statute has been, to censure it for taking kernels of wheat as the natural standard of weights; with the very obvious remark that the wheat of different seasons and of different fields, and often even of the same field and the same sea-

son, is different. But the statute is chargeable with no
such uncertainty. The statute merely describes how the
standard measure of the exchequer, by the consent of the
whole realm of England, was made. The article, for
which of all others the measure was most wanted, was
wheat; and a measure was wanted which should give it,
as far as it was practicable, in number, weight, and measure. It took, therefore, thirty-two kernels of *average*
wheat from the middle of the ear, and found them equal
in weight to the silver penny sterling, new from the mint,
round and without clipping. It then drops the numeration of wheat; but proceeds to declare that twenty such
pence make an ounce, twelve ounces one pound, and eight
pounds a gallon of *wine,* and eight *gallons of wine* a London bushel, which is the eighth part of a quarter. It must
be observed here, that it was not the *measure* but the
weight of wine, which was used to form the standard
bushel. It was not eight wine gallons, but eight gallons
of wine. The bushel, therefore, filled with wheat, was a
measure which, in the scales, would exactly balance a keg
containing eight gallons of wine, deducting the tare of
both the vessels. Now, the eighth part of this bushel, or
the ale gallon, would be a vessel, not of the same cubic contents as the wine gallon, but of the same proportion to it
as the weight of wheat bears to the weight of wine;
the proportion between the commercial and nummulary
weights of the Greeks; the proportion between our avoirdupois and troy pounds.

But neither the present avoirdupois, nor troy weights,
were then the standard weights of England. The key-
stone to the whole fabric of the system of 1266 was the

weight of the silver penny *sterling.* This penny was the two hundred and fortieth part of the tower pound; the sterling or easterling pound which had been used at the mint for centuries before the Conquest, and which continued to be used for the coinage of money till the eighteenth year of Henry the Eighth, 1527, when the troy pound was substituted in its stead. The tower or easterling pound weighed three-quarters of an ounce troy less than the troy pound, and was consequently in the proportion to it of 15 to 16. Its penny, or two hundred and fortieth part, weighed, therefore, $22\frac{1}{2}$ grains troy; and that was the weight of the thirty-two kernels of wheat from the middle of the ear, which, according to the statute of 1266, had been taken to form the standard measure of wheat for the whole realm of England. It is also to be remembered, that the eight twelve-ounce pounds of wheat, which made the gallon of wine, produced a measure which contained nearly ten of the same pounds of wine. The commercial pound, by which wine and most other articles were weighed, was then of fifteen ounces. This is apparent from the treatise of weights and measures of 1304, which repeats the composition of measures declared in the statute of 1266, with a variation of expressions, entirely decisive of its meaning. It says that "by the ordinance of the whole realm of England, the measure of the king was made, that is to say: that the penny called sterling, round, and without clipping, shall weigh thirty-two grains of wheat in the middle of the ear. And the ounce shall weigh twenty pence; and twelve ounces make the *London* pound; and eight pounds of *wheat* make a gallon; and eight gallons make the London bushel." It then proceeds

to enumerate a multitude of other articles, sold by weight
or by numbers, such as lead, wool, cheese, spices, hides, and
various kinds of fish; and, after mentioning nominal hundreds, consisting of 108 and 120, finally adds, "it is to be
known that every pound of money and of medicines consists only of twenty shillings weight; but the pound of *all*
other things consists of twenty-five shillings. The ounce
of medicines consists of twenty pence, and the pound contains twelve ounces; but, in other things, the pound contains fifteen ounces, and, in both cases, the ounce is of the
weight of twenty pence."

Wine and wheat therefore were both among the articles
of which the pound consisted of fifteen ounces. By the
statute of 1266, the gallon of wine contained eight such
pounds of wine. By the statute of 1304, the gallon (for
ale) contained eight such pounds of wheat; and the weight
of wine contained in eight such wine gallons, and the
weight of wheat contained in eight such corn or ale gallons, was equally the measure of the bushel.

The wine, to which the statute of 1266, and many subsequent English statutes exclusively refer, was the wine of
Gascoign, a province at that, and for a long period, under
the dominion of the English kings, the same sort of wine
which now goes under the name of Claret, or Bordeaux.
Its specific gravity is to that of distilled water as 9,935 to
10,000, and its weight is of 250 grains troy weight to the
cubic inch.

With these data we are enabled, accurately, to ascertain
the dimensions and contents of the bushel, the ale gallon,
and wine gallon, of 1266. The silver penny, called the
sterling, to which 32 kernels of wheat were equiponderant,

was equal to 22½ grains troy. Its pound of twelve ounces was equivalent to 5,400 grains troy. The pound of fifteen ounces, by which wheat and wine were weighed, was equal to 6,750 grains troy. Eight such pounds were equal to 54,000 grains troy, which divided by 250, the number of grains troy, weighed by a cubic inch of Bordeaux wine, gives a wine gallon of 216 cubic inches.

There is no standard wine gallon of that age extant in England; but the weights and measures of England were established by law in Ireland as early as the year 1351: and by the act called Poyning's law, of 10 Henry VII. (1493), all the then existing statutes of England, relating to weights and measures, were made applicable to Ireland. The changes since effected in England have not extended to Ireland; at least in relation to the measure of wine. The standard Irish wine gallon at this day is of 217.6 cubic inches; a difference almost imperceptible in the quantity of the gallon, from the legal standard of 1266, and the cause of which must have been this.

There was another law, probably of date more ancient than the year 1266, in which the measure of the wine gallon was fixed by a different process.

A statute of the year 1423, the second of Henry the Sixth, ch. ii., declares that, "*in old time* it was ordained, and lawfully used, that tuns, pipes, tertians, hogsheads of Gascoigne wine, barrels of herring and of eels, and butts of salmon, coming by way of merchandise into the land, out of strange countries, and also made in the same land, should be of certain measure: that is to say: the tun of wine 252 gallons, the pipe 126 gallons, the tertian 84 gallons, the hogshead 63 gallons, the barrel of herring and of eels 30 gallons, fully packed, the butt of salmon 84 gallons, fully packed, etc. ; but that of late, by device and subtlety, such vessels have been of much less measure, to the great deceit and loss of the king and his people, whereof special remedy was prayed in the parliament." It then proceeds to re-enact that no man shall make in England vessels for those purposes, or bring wine, etc., into England in vessels of other dimensions than those thus prescribed, upon penalty of forfeiture.

The ordinance of *old time*, referred to in this act, is not now among the statutes at large, and is therefore probably of more ancient date than the Magna Charta of 1225. As it regulated the size of casks, which, in the nature of the thing, were to be made in the country whence the wine was imported, it seems likely to have originated when Gascoign was under English dominion, and when the law of Bordeaux could be accommodated to the assize of the English ton. This assize of the ton is in its nature connected with the trade of the cooper, with the assize of hoops and staves, with the art of the ship-builder, and with the whole science of hydraulics and of navigation. The measure and form of the ton must be accommodated to the character of the substance which it is to contain, and to the convenience and safety of its conveyance by sea. It must be adapted for stowage to the necessary form of the ship; to the volatile property of fluids; to the concussions of tempestuous elements. It is in the composition of the ton that the natural connection between the weight of water, and cubic linear measure, first presents itself. The burden of the ship is the weight of tonnage which it can bear afloat upon the waves; that weight is equal to the weight of water which it displaces; the measure of the ship must be taken by the builder in linear measure. Now eighty of the old easterling tower pounds make 432,000 grains troy weight, which, divided again by 250, the number of troy grains to a cubic inch of Bordeaux wine, give 1728 cubic inches, precisely the dimensions of an English cubic foot, one-eighth part of which makes again the gallon of 216 cubic inches. And here we discover, again, the quadrantal or amphora of the Romans, the cubic foot containing 80 pounds of wine.

That the assize of the ton, which in 1423 was *of old time*, was equally well known and established in 1353, appears from a statute of that date, 27 Edward III., ch. 8, directing that all wines, red and white, should be gauged by the king's gaugers, and that in case less should be found in the tun or pipe than ought to be of right, *after the assize of the tun*, the value of as much as lacked should be allowed and deducted in payment.

The casks of Bordeaux wine were then and still are made for stowage in such manner that four hogsheads occupy one ton of shipping. The ton was of thirty-two cubic feet by measure, and of 2,016 English pounds, of fifteen ounces to the pound, in weight; equal to 2,560 of the easterling tower pound.

In comparing together the wine gallon as prescribed by the statute of 1266 and that derived from the assize of the tun, we find the former in the ascending ratio, beginning with the kernel of wheat and multiplying: the latter is formed in the descending ratio, beginning at the tun and dividing. In one process, the gallon is formed by weight; in the other, by measure. The hogshead of wine was the measure corresponding to the quarter of wheat: but there was a difference of eight pounds in their weight. The hogshead of wine weighed 504 and the quarter of wheat 512 pounds, of 15 ounces. The wine gallon of 216 cubic inches, prescribed by the statute of 1266, was thus an exact eighth part of the English cubic foot of 1728 inches.

The wine gallon therefore is the *congius* of the Romans, weighing ten nummulary and eight commercial pounds, and measuring exactly the eighth part of a cubic foot.

But the gallon of 216 cubic inches, the eighth part of the cubic foot, was derived originally from a measure of *water*, and was an aliquot part of the ton of shipping. The wine gallon of 1266 was made of eight easterling pounds of wheat; and, therefore, contained of water eight corresponding commercial pounds. But if the gallon of water, weighing eight pounds, was of 216 solid inches, the gallon of Gascoign wine, to be of the same weight, would be of 217.6 solid inches, the precise contents of the standard Irish gallon to this day: and the specific gravity of that wine being to that of wheat as 143 to 175, the corn gallon, balanced by this Irish gallon of 217.6 inches, must be of 266.17 cubic inches. The Rumford corn gallon of the year 1228, examined by the committee of the House of Commons in 1758, was found to be of 266.25 cubic inches. The Irish wine gallon and the Rumford corn gallon of 1228 were both made, with an accuracy which all the refinements of art of the present-age could scarcely surpass, from the standard measure made, as the statute of 1266 declares, by the consent of the whole realm, and precisely in the manner therein described.

But as the hogshead, measuring eight cubic feet, was

required by the assize of the tun to contain only sixty-three gallons of wine, it followed of course that the gallon thus composed was of 219.43 cubic inches; and as the weight of eight such gallons of wine was to form the bushel, the proportion of the weight of wine being to that of wheat as 143 to 175, the bushel would be of 2148.25 cubic inches, which is within two inches of the Winchester bushel.

This system of weights and measures has been, by many of the modern English writers on the subject, supposed to have been *established* by the statute of 1266. But, upon the face of the statute itself, it is a mere exemplification of ancient ordinances. The coincidences in its composition with those of the ancient Romans, proved by the letter of the Silian law, and by the still existing congius of Vespasian; with those of the Greeks, as described by Galen, and as shown by the proportions between their scale weight and their metrical weight; and with that of the Hebrews, as described in the prophecy of Ezekiel; show that its origin is traceable to Egypt and Babylon, and there vanishes in the darkness of antiquity. As founded upon the identity of nummulary weights and silver coins, and upon the relative proportion between the gravity and extension of the first articles of human traffic, corn and wine, it is supposed to have originated in the nature and relations of social man, and of things.

It has been said, that the first inroad upon this system in England was made by Edward the First himself, by destroying the identity between the money weight and the silver coin. From the time of the Norman Conquest, and long before, that is, for a space of more than three centuries, the tower easterling or sterling pound had been coined into twenty shillings, or two hundred and forty of those silver *pennies*, each of which weighed thirty-two kernels of wheat from the middle of the ear. Edward the First, in the year 1328, coined the same pound into two hundred and forty-three pennies of the same standard alloy. From the moment of that coinage, the penny called a sterling, however round, however unclipped, had lost the *sterling* weight, though it still retained the name. This debasement of the coin, once commenced, was repeated by successive sovereigns, till, in the reign of Edward the Third, the pound was coined into twenty-five shillings, or three hundred pennies. The silver penny then weighed only 25⅕ kernels of that wheat of which the penny of 1266 weighed 32. It is probable that, in reducing the weight of their coins, none of those sovereigns were aware that they were taking away the standard of all the weights and of all the vessels of measure, liquid and dry, throughout the kingdom; but so it was. It destroyed all the symmetry of the system. It has been further affected by the introduction of the troy and avoirdupois weights.

The standard measures of the exchequer had been made by the rules set forth in the statutes of 1266 and 1304. These standards were kept in the royal exchequer. In process of time the standards themselves fell into decay, and called for renovation. In the year 1494, shortly after the termination of the long and sanguinary wars between the houses of York and Lancaster, Henry the Seventh, in the tenth year of his reign, undertook to furnish forty-three of the principal cities of the kingdom with new copies of all the standard weights and measures then in the exchequer. They were accordingly made and delivered to the representatives in parliament of the respective counties; but it was soon discovered that they were all defective, and not made according to the laws of the land. From what cause this had arisen does not appear; but that the laws of the land to which they referred, namely, the statutes of 1266 and 1304, were and continued to be entirely misunderstood, is abundantly apparent from the statute which was made the very next session of parliament, 1496, to remedy the evil.

This act, after reciting the extraordinary attention of the king in having made at his great charge and cost, and having distributed, all those county standards of weights and measures, according to the old standards in the treasury; and after stating the disappointment which had ensued, upon the discovery of more diligent examination that they were all defective and not made according to the old laws and statutes, proceeds to ordain, that the measure of a bushel contain eight gallons of wheat, that every gallon contain eight pounds of wheat, *troy* weight, and every pound contain twelve ounces of troy weight, and every ounce contain twenty *sterlings*, and every sterling be of the weight of thirty-two corns of wheat that grew in the midst of the ear of wheat, *according to the old laws of the land;* and the new standard gallon, after the said assize, was to be made to remain in the king's treasury forever. All the weights and measures, which had been sent by the act of the former year throughout England, were directed to be returned; others, conformable to the new standard, were to be made from them and sent back; after which, all weights and measures were to be made conformably to them.

It is from the terms of this statute that many of the English writers have concluded that the kernel of wheat was the original standard of English weights. It is by this statute made the standard of troy weight; but it was not so according to the *old laws of the land.* It was not so in the measure declared in 1266 and 1304 to have been made by the consent of the whole realm of England. To prove this, it is only necessary to compare the statutes together.

The **two first** declare, that an English *penny, called a sterling,* round and without any clipping, will weigh thirty-two corns of wheat from the midst of the ear. That penny was the two hundred and fortieth part of the old tower pound, and was one-sixteenth lighter **than** troy weight. The weight of that penny in 1266 is therefore now known, but appears not to have been known to the parliament of 1496. For the tower pound was then coined into thirty-seven shillings and six pence sterling, and, consequently, the penny called a sterling, instead of *then* weighing thirty-two grains of the wheat, which it weighed in 1304, would have weighed only seventeen of the same grains.

The term *penny,* therefore, is dropped in the act of 1496, but the term sterling is retained, and improperly applied to the penny *weight* troy. The penny of 1266 was both weight and coin. In 1496, the penny had ceased to be a coin, and the penny sterling, which was yet money, weighed little more than half what it had weighed till after 1304. The penny weight troy was never called *a sterling,* anywhere, or at any time, but in this act of 1496. It was neither the weight of the old tower standard, nor was it the penny sterling of Henry the Seventh's own coinage.

The statute of 1496 inverts the order of the old statutes; it is not a *composition,* but an *analysis,* of measures. It begins with the bushel, and descends to the kernel. The act of 1266, to make the weight, number, and measure, of corn, money, and wine, begins with the kernel, and ascends by steps to the weight of coin; thence to the measure of wine, by the weight of corn; thence to the measure of corn, by the weight of wine. The mere process of the composition establishes the proportional measures. The statute of 1496 destroys the proportion altogether. It says that every gallon shall contain eight pounds of wheat *troy weight,* and every pound twelve ounces of troy weight. It substitutes, therefore, instead of the *weight* of the gallon of wine, prescribed by the statute of 1266, the *measure* of the wine gallon, for the eighth part of the bushel. The gallon, established by this act of 1496, is the gallon of two hundred and twenty-four cubic inches; the Guildhall gallon, which in 1688 was found by the commissioners of the excise to be of that capacity. It contains eight pounds troy weight of wheat, and, consequently, eight pounds avoirdupois of Bordeaux wine, of 250 grains troy to the cubic inch. Its bushel would contain seventeen hundred and ninety-two cubic inches; but if such a bushel ever was made, as the act required, it never was used as a standard. It must have been found to fall too far short of the old standards still existing; and the real standard bushels of Henry the Seventh, in the exchequer, instead of being made according to the process prescribed in his law of 1496, must have been copied from the older standard bushels then existing.

The gallon of two hundred and thirty-one inches was also a gallon made under the statute of 1496. But the wheat is of that kind thirty-two grains of which equipoise the penny of the old tower pound; while the wheat that forms the gallon of two hundred and twenty-four inches, is that of which thirty-two kernels weigh a penny weight troy. The weight of the corn in both gallons would be the same; but that, of which each kernel upon the average would be one-sixteenth heavier than those of the other would, by the combined proportion of gravity and numbers, occupy one thirty-second less of space. This is precisely the difference between the gallons of two hundred and twenty-four and two hundred and thirty-one solid inches.

The debasement of the coin had destroyed its original identity with the money weight. The substitution of *troy* weight, instead of the old easterling pound, for the composition of the gallon, destroyed the coincidence between the water gallon, derived from the ton, the eighth part of the cubic foot, and the wine gallon, containing eight money pounds of wheat. The wine gallon of two hundred and twenty-four, or two hundred and thirty-one cubic inches, no longer bore the same proportion to the cubic foot of water; one consequence of which was, that the hogshead of Bordeaux wine, which the law required to contain sixty-three gallons, no longer contained that number of English gallons; but, from that day to this, has contained from fifty-nine to sixty-one. It still contains at least sixty-three Irish gallons.

The act of 12 Henry VII. (1496), intended, upon the face of it, to be a mere repetition of the statutes of 1266, and 1304, was thus a total subversion of them. It was founded upon two mistakes; the first, a supposition that the penny sterling, described in those statutes, was the penny weight troy; and the second, a belief that it was the *measure*, and not the *weight*, of eight gallons of wine, which constituted the bushel. The causes of these mistakes were, first, that, in the lapse of two centuries, a great portion of which had been a period of calamity and civil war, the successive debasements of the coin had reduced the penny sterling to about half its weight as it was when made the standard of comparison with thirty-two kernels of wheat; and finding that the penny sterling of their own time, if used to make the new standard bushel, would reduce its size by nearly one-half, which had perhaps been the mistake upon which the act of 1494 had been made, they must hastily have concluded that it was the penny weight troy, which was intended by the old statutes. The second was a misapplication of the term *gallon*, which, in its original meaning, and in its popular signification to this day, is exclusively a measure of liquids, and not of dry substances. In the statute of 1266, it is expressly called the gallon of *wine*. In the act of 1304, it was called the gallon, without addition, but meaning the same wine gallon. The measure for corn was the bushel; and its subdivisions were the peck, pottle, and pint. The eighth part of the *measure* of a bushel was first called a *gallon*, because it was used as the measure of the ordinary liquids brewed from grain, beer and ale. There never was properly any corn gallon; and the term was misapplied even to denote the measure of beer. Being a vessel of different dimensions, it ought to have had a different **name**; and that alone would have prevented them from **ever** being mistaken the one for the other.

The parliament of 1496 were seduced by those expressions, so often re-echoed from year to year, and from century to century, that there should be but one weight and one measure throughout the land. They mistook the uniformity of proportion for the uniformity of identity. They construed the threefold "one measure of wine, and one measure **of** ale, and one measure of corn," ordained in Magna Charta, as if it meant that those three one measures should **be the** same. That these mistakes should be made is not surprising, when we consider that, in 1496, **the art of printing** was but in the cradle; that no collection **of** the statutes had ever been printed; that the languages **in** which the statutes of 1266 and 1304 had been enacted, the Norman French and the barbarous Latin of the thirteenth century, were no longer in use, at least in parliament; that the very records by which the weight of the penny sterling in 1266 might have been ascertained, were, perhaps, not known to exist. But it is not so easy to explain how they could mistake the penny of the old easterling pound, which was still, and continued for forty years after to be, used at the mint for coining money for the penny troy weight.

We have seen that, in 1304, the easterling pound of twelve ounces was the money pound, and that the corresponding commercial weight was a pound of fifteen of the same ounces. These were the result of a rough and inaccurately settled proportion between the specific gravity of wheat and wine, or wheat and water. Mr. Jefferson has justly remarked, that the difference between the specific gravity of wine and of water is so small that it may safely, as between buyer and seller, be disregarded. And it was disregarded by those two acts of parliament, one of which made the wine gallon an eighth part of the cubic foot of water, while the other made it equiponderant to eight easterling pounds of wheat. So the proportion of the two pounds of twelve and fifteen of the same ounces was, upon a rough estimate, that the proportional weight of wheat and wine was as four to five, or as fourteen to seventeen and a half; and it was afterward assumed as of fourteen to seventeen. But if trade, and even legislation, may safely neglect small quantities, nature is no such accountant of more or less. It has been shown that the water gallon of eight easterling twelve-ounce pounds of wheat corresponds with the gallon of two hundred and sixteen cubic inches, the eighth part of the cubic foot; but, that when a Bordeaux *wine* gallon is to be made, containing the same eight pounds of wheat, it produces a gallon, not of 216, but of 217.6 cubic inches. In the mode of forming the gallon and bushel, described in the act of 1266, it is not the loose calculations of man, but the unerring hand of nature, that establishes the proportions. The vessel that would hold eight money pounds of wheat was the wine gallon. The vessel that would balance, filled with wheat, eight gallons of that wine, was the bushel. Then, if half a peck of this bushel was taken for a beer gallon, its proportion to the wine gallon would not be of fifteen to twelve, nor of seventeen to fourteen, but of one hundred and seventy-five to one hundred and forty-three.

When the avoirdupois or the troy weights were first introduced into England has been a subject of controversy among the English writers, and has not been ascertained. The names of both indicate a French origin : but that no new weight or measure was brought in by the Norman Conqueror is certain, and the statute of weights and measures of 1304 proves that neither of them were *then* recognized by law. One of the most learned writers upon the coins,* says that troy weights were first established by this statute of Henry VII. of 1496; that it was owing to the *intercursus magnus*, or great treaty of commerce concluded between England and Flanders the year before; that the Flemish pound was adopted as a compliment to the Duchess of Burgundy, and for the mutual convenience of all their payments, which would then be adjusted by the same pound.

This conjecture is ingenious, but not well founded. The statute of 1496 did, in fact, legitimate troy weight for the composition of the gallon and the bushel, but it professed, and intended to introduce, no new weight or measure. Its purpose was to re-enact the composition of weights and measures of 1266. It was a legislative error intended to correct another error committed at the last preceding session of parliament in 1494, before the intercursus magnus was concluded. Instead of correcting the error, it rendered it irreparable. It was so far from correcting the error, that, although a standard wine gallon was made under this statute, which was the Guildhall wine gallon of two hundred and twenty-four inches, there never was a standard bushel made by the rule prescribed in this statute: and if there had been, its cubic contents would

---

* Clarke.

have been not one inch more or less than seventeen hundred and ninety-two.

The troy weight was never used by Henry VII. at the mint at all. He made a wine gallon by it, because the difference between a gallon of 217.6 inches, which was the old standard, and one of 224, which was made by his troy weights, was not large enough to make its incorrectness apparent. It was scarcely the difference of a small wine glass upon a gallon: and, as it was a difference of excess over the contents of the old standard, it might naturally be attributed to the decays or inaccuracy of that. He ordained that a bushel should be made by it; but a bushel made from the *measure* of his wine gallon, a bushel of seventeen hundred and ninety-two inches, would have contained at least three hundred and thirty inches, nearly twelve pounds in weight less than any of the old standards. This would have been found a difference utterly intolerable. It would have been necessary to recall and break up the new standards a second time, and to acknowledge a second error greater than the first. So the statute, so far as related to the composition of the bushel, was suffered to slumber upon the rolls; the old standard bushels were still retained; and new ones were also made, not by the troy, but by the avoirdupois pound of wheat: and hence it is that standard bushels of Henry the Seventh exist at the exchequer, one of 2,124 inches, which is the old standard, and one of 2,146 inches, which is the Winchester bushel, and, at the same time, corn gallons of 272 inches.

That the Troy weight was not introduced into England by Henry the Seventh is further proved by two statutes;

one of 1414, 2 Henry V., ch. 4, and one of 1423, 2 Henry VI., st. 2, ch. 4: in the first of which it ordained that the goldsmiths should give no silver worse than of the allay of the English sterling, and that they take for a pound of *troy* gilt but forty-six shillings and eight pence at the most; and, in the second, that silver, not coined, in plate, piece, or in mass, being of as good allay as the sterling, should not be sold for more than thirty shillings the pound troy, besides the fashion, because the same was of no more value at the coin than thirty-two shillings. The tower easterling pound was at that time coined into thirty shillings, and the value of the troy pound of the same alloy was, of course, thirty-two.

From these two statutes it is apparent that, nearly a century before Henry the Seventh, the troy pound was used by the goldsmiths, who were the bankers of that age, and were foreigners, for weighing bullion and plate; and that the proportion of the troy pound to the tower money pound was as sixteen to fifteen.

That the troy pound, though adopted by Henry the Seventh for the composition of the bushel and gallon, was not introduced by him at the mint, appears equally clear. About the middle of the last century, Martin Folkes published his tables of English coins, in which he cited a verdict remaining in the exchequer, dated 30th October, 1527, 18 Henry VIII., in which are the following words: "And whereas, heretofore, the merchant paid for coynage of every *pound towre* of fyne gold, weighing xi. oz. quarter troye, 2*s.* 6*d.* Nowe it is determined by the King's highness, and his said councelle, that the aforesaid pounde towre shall be no more used and occupied; but all manner of golde and sylver shall be wayed by the pounde troye, which maketh xii. oz. troye, which excedith the pounde towre in weight three quarters of the ounce."—[*Clarke*, p. 15.]

A French record of much earlier date, from the register of the Chamber of Accounts at Paris, cited also by Folkes, shows that, early in the fourteenth century, there were among the weights in common use in France two marks, of different gravity, one of troy, and the other of *Rochelle*, in the same proportion to each other, and that the last was called the mark of *England*.

The Rochelle and easterling pound was therefore the same; and that was the pound, eight of which in spring-water were contained in the eighth part of the cubic foot, and formed the gallon of 216 cubic inches.

This proportion, as has been observed, was totally destroyed by the substitution of the troy pound by Henry the Seventh, in 1496, instead of the Rochelle pound, for the composition of the bushel and gallon.

As, by the treatise of weights and measures of 1304, only two weights are mentioned, by which it asserts that all things were weighed, this tower pound of twelve ounces, and the corresponding commercial pound of fifteen of the same ounces, it is clear that the troy weight was then unknown, or at least not used in England. But this reign of Edward the First was also the period when the foreign commerce of England began to flourish. In 1296, the famous mercantile society, called the Merchant Adventurers, had its first origin; and another society of natives of Lombardy, the great merchants of that age, about the same time established themselves in England, under the protection of a special charter of privileges from

Edward.   These Lombards soon became the goldsmiths and Bankers in England.   [*Hume,* vol. ii. p. 330, ch. 13.] In the year 1354, the balance of exports above the imports was of more than 250,000 pounds; and as the balances of that age could be paid for only in specie, whenever the balance was in favor of England it must have brought much foreign money into the kingdom.   The pound of the goldsmiths and bankers was the troy weight, and by them, **there can** be little doubt, it was first introduced. The pound of fifteen ounces troy must have been introduced **at** the same time, by an accommodation of that **weight to** the old English rule—that when bullion and **drugs were** weighed by a pound of twelve ounces, **all other things were** weighed by **a** pound of fifteen of the same ounces.   This pound **of fifteen ounces, or** 7,200 grains troy, has never been **recognized in** England by law; but, in many parts of England, it has been used under the name of the merchant's weight: and eight such pounds of wheat form precisely the gallon of 280 cubic inches, of which **the** standard quart in the exchequer is the fourth part.

The time and occasion of the introduction of the avoirdupois pound into England is no better known than that of the troy weight.   But it may be inferred, from the ancient statutes, that it was brought in by the same foreign merchants, **with** the troy, and as the corresponding weight **to that as** the weight for bullion **and** pharmacy. The first time that the term avoirdupois is used in the English statute-book, is in the 9th of Edward III., stat. 1, ch. 1 (1335), the very statute which authorizes merchant *strangers* to buy or sell corn, wine, *avoirdupois,* flesh, fish, and all other provisions, and victuals, wool, cloths, wares, merchandises, and all other vendible articles in any part of England.

Eighteen years afterward, in the celebrated statute staple of 1353 [27 Edward III., ch. 10], is the following provision: "Forasmuch as we have heard that some merchants purchase *avoirdupois*, woollens, and other merchandises by one weight and sell by another, and also make deceivable diminutions upon the weight, and also use false measures and yards, to the great deceit of us and of all the commonalty and of honest merchants: We therefore will and establish, that one weight, one measure, and one yard, be throughout the land, as well without the staple as within, and that woollens, and all manner of *avoirdupois*, shall be weighed by balances," &c.

In these two statutes, the term avoirdupois manifestly refers, not to the weight, but to the article weighed. It means all *weighable* articles, in contradistinction to articles sold by measure or by tale; and the *one weight*, meant by the statute staple, is the easterling pound of fifteen ounces, mentioned in the statute of weights and measures of 1304. Money and bullion were not included among these *weighable* articles, because they had a speciable weight of their own, and because money was current by *tale*. Grain and liquids were not weighable articles, because they were bought and sold by *measure*. Hence arose naturally the practice of calling all articles bought and sold by weight in the traffic of these merchant strangers, articles *having weight*, or weighable.

That this is the meaning of the term avoirdupois is also demonstrated by an act of 1429 (8 Henry VI., ch. 5), which, reciting these regulations of Edward the Third, expressly says, that they require woollens, and all manner of *weigh-able* things [toutz manerz des choses poisablez] bought or sold, to be weighed by even balance, with weights scaled according to the standard of the exchequer.

The terms *avoirdupois*, and *choses poisablez*, were therefore synonymous. But the merchant strangers had a weight of their own, the corresponding commercial weight proportional to their pound troy. This was the weight now called the avoirdupois pound, of sixteen ounces, but the ounce of which was not the same as that of the troy weight. The standard easterling pound of fifteen ounces at the exchequer weighed 6,750 grains troy. The avoirdupois pound of the merchant strangers weighed 7,000. The difference between them was but of about half an ounce, and one sees instantly what temptations and opportunities this slight difference furnished to those unprincipled merchants, of whom the statute staple complains, of buying by their own foreign larger weight, and selling by the weight of the exchequer.

The statute staple of 1353, and the act of 2 Henry VI., 1423, are both evidences of the conflict between the mint and exchequer easterling pounds on one side, and the troy and avoirdupois weights of the merchant strangers on the other. In this struggle the latter ultimately prevailed, and completely rooted out the old English weights. The troy weight, being adopted by Henry the Seventh, in 1496, for the composition of the bushel and gallon, and by Henry the Eighth, in 1527, for making money at the mint, the avoirdupois pound came in as the corresponding commercial pound; and a statute of 24 Henry VIII., ch. 3, 1532, directs, that beef, pork, mutton, and veal, shall be sold by weight, *called* haverdupois; the very use of which expression, *called* haverdupois, indicates that it was a denomination, as applied to the weight, of recent origin, and that the weight itself had not been long in general use for any purpose.

A statute of the preceding year, 23 Henry VIII., ch. 4, 1531, s. 13, had directed, that every cooper should make every barrel for ale "*according to the assize specified in the treatise called Compositio Mensurarum* [the statute of 1304]; that is to say, every barrel for ale shall contain thirty-two gallons of the said assize, or above, of the which eight gallons make the common bushel to be used in this realm of England," etc.

By this statute the ale gallon was expressly declared to be the eighth part of the *measure* of the bushel. Now, it has been proved that, by the *Compositio Mensurarum*, the bushel was a measure containing of wheat the *weight* of eight gallons of wine. The eighth part of this *measure*, therefore, being the ale gallon, *must* bear the same proportion to the wine gallon as the specific gravity of wheat bears to that of wine; and the wine gallon of 231 inches not having been made by the rule of the *Compositio Mensurarum*, but by the troy weight of the statute of 1496, that is to say, weighing eight troy pounds of the wheat thirty-two kernels of which were equiponderant to the penny sterling of 1266, the bushel, to balance eight such gallons of wine, must of necessity contain sixty-four avoirdupois pounds of wheat, and measure 2,256 cubic inches. The eighth part of this measure is the gallon of 282 inches, which is to this day the standard ale and beer gallon of the British exchequer and of these United States.

And thus we have seen that all the varieties of standard gallons and bushels which have been found—from the Irish gallon of 217.6 cubic inches, to the ale gallon of 282, and from the ordained, but never made, bushel of 1792 inches, prescribed by Henry the Seventh's act of 1496, to the bushel of 2,256 inches, of which the ale gallon is the eighth part—are distinctly traceable to the inconsistency of human laws, and the consistency of the laws of nature.

The statute of 1496 changed the contents of both the gallons and of the bushel, without intending it. For, although the bushel of 1792 inches was never made, or, at least, never deposited as a standard at the exchequer, yet new standard bushels were made from the new wine gallon by the rule of the *Compositio Mensurarum*, and they produced the bushel of 2224 of Henry the Seventh, still extant at the exchequer.

The *Winchester* bushel of the exchequer, however, was not thus made. It was found, in the year 1696, to contain 2145.6 cubic inches of spring-water. Its ale gallon, therefore, by the statute of 1531, and the *Compositio Mensurarum*, would have been of 268.2, and its wine gallon of 219.2, cubic inches. Its difference from the proportions of the Irish wine gallon and the Rumford corn gallon is so slight, that there can be no doubt it was copied from a model made by the statute of 1266.

That the capacity of the wine gallon, although it was very essentially changed, was not intended or understood to be changed by the statute of 1496, is proved by the statute of 28 Henry VIII., ch. 14, 1536; which re-enacts the old statutes requiring that the tun of wine should contain 252 gallons, and all other casks of the same pro-

portion, including the hogshead of sixty-three gallons.
Now, the hogshead which contained sixty-three gallons of
217.6 cubic inches, could contain no more than sixty-one
and a quarter gallons of 224 inches, nor more than fifty-
nine and one-third gallons of 231 inches. The ordinary
Bordeaux hogshead contains from fifty-nine to sixty gallons; and the size of this cask, being formed of a certain
number of staves of settled dimensions, and made by the
coopers in particular forms, has passed down from age to
age without alteration; while the laws of England, and
those of several of the United States, have required that
it should contain sixty-three gallons of 231 cubic inches,
because, five hundred years ago, the laws required it to
contain sixty-three gallons of 219.5 inches.

In the reign of Elizabeth, the change, which had been
effected in the wine gallon by the act of 1496, was discovered in its effects upon another branch of trade; but
the cause of the change appears not to have been perceived. The statute of 13 Elizabeth, ch. 11 (1570), recites,
that the people employed in the herring fishery "had,
time out of mind, used to pack their herring in barrels
containing *about* thirty-two gallons of usual *wine* measure,
which assize had always been gauged and allowed in the
city of London, yet the measure had lately been quarrelled
at by certain informers, for not containing thirty-two gallons by the old measure of standard, *which they never did*,
though peradventure *the extremity of old statutes in words,
by some men's construction*, might be stretched to require
so much." It then enacts, "that thirty-two gallons wine
measure, which is *about* twenty-eight gallons by old standard, shall be the lawful assize of herring barrels, any old

statute to the contrary notwithstanding." This was cutting the gordian knot. The wine gallon here referred to was the gallon of 231 inches, made by the troy weight wheat of Henry the Seventh. The *old standard* is the corn gallon of 1266, which, according to the Rumford quart examined by the committee of the House of Commons in 1758, was of 264.8, or, according to the Rumford gallon, was of 266.25 cubic inches. Now twenty-eight gallons of 264 inches are of precisely the same capacity as thirty-two gallons of 231. But, as the wine gallon at Guildhall, though it showed 231 inches by the gauge, did, in fact, contain seven inches less, and as the herring barrels were gauged according to the Guildhall gallon, they would have fallen short nearly one gallon in the barrel of twenty-eight gallons by the *old standard:* and the act, the object of which was to rescue the herring fishers from the fangs of informers, is cautious not to tie them down to too close a measure. The old statutes, the construction of which the act professes to consider as doubtful, are not named; but the act of 23 Henry VIII., 1531, *must* have been that upon which the informers had quarrelled at the assize of the barrels used by the herring fishers.

That act requires that the coopers should make barrels of thirty-two gallons for ale, according to the assize of the Compositio Mensurarum—gallons, eight of which make the common bushel. Now, the act of 1496 had expressly directed that *every* gallon should contain eight pounds of wheat *troy* weight, and that the bushel should contain eight such gallons of wheat. But this law, so far as it prescribed a new bushel, had never been executed: the old standard bushels remained. So that the statute for the coopers of 1531 was on the side of the informers, and the statute of weights and measures of 1496 was on the side of the herring fishers. Parliament found no expedient for the difficulty but to declare the usual existing size of the herring barrels lawful, and to set all the old statutes aside, with a non obstante.

If the parliament of 1496, contrary to their avowed intention, did actually change the capacity of the wine and corn gallons, and did ordain a much greater change of the capacity of the bushel, these varieties, effected by the law, while they were unknown to the legislators, were still less likely to come to the general knowledge of the people. The eagle eyes of informers would occasionally discover that the measures of the people fell short of the standards of the law: but the people took the standards as they came, and used the measures which they and their forefathers, time out of mind, had employed.

The Restoration of 1660, after the convulsions of a civil war, formed a new æra, not only in the political history of England, but in that of their vessels of capacity. It was then that a new system of taxation commenced by the excise upon liquors. About the same time, also, commenced a new æra in the philosophical and scientific pursuits of the English nation, by the establishment of the Royal Society. Both these events were destined to have an important influence upon the history of English weights and measures.

The excise was a duty levied, by the gallon, upon malt liquors and upon wines. The malt liquors were measured by the standard gallon at the treasury, made according to the cooperage act of 1531, by the rule of the *Compositio*

Mensurarum, applied to the troy weight wine gallon of the statute of 1496. This gallon, therefore, was neither the wine gallon of 1496, nor the eighth part of the old standard bushel, nor of the Winchester bushel, but the gallon which, if filled with wheat of the troy weight specified in the statute of 1496, would balance the wine gallon of 231 inches; it was, therefore, a gallon of 282 cubic inches. The wine was measured by the gauge of the wine gallon at Guildhall.

Taxation and philosophy now began to speculate, at the same time, upon the weights and measures of England. In 1685, the weight of a cubic foot of spring-water was found, by an experiment made at Oxford, to be precisely 1000 ounces avoirdupois; and, in 1696, the Winchester bushel to be of 2145.6 cubic inches, and to contain, also, 1000 ounces avoirdupois weight of wheat. Yet so totally lost were all the traces of the old easterling pounds of twelve and fifteen ounces, that this coincidence between the cubic foot of water, and the 1000 ounces avoirdupois, gave an erroneous direction to further inquiry; for the real original connection between the cubic foot and the English bushel was not formed by avoirdupois weights and water, but by the easterling pound of twelve and fifteen ounces and Gascoign wine. It was the principle of the quadrantal and congius of the Romans, applied to the foot and the nummulary pound of the Greeks; the measure which, by containing eight pounds of wheat, was intended to contain, at the same time, ten of the same pounds of wine.

In the year 1688, the commissioners of the excise instituted an inquiry why beer and ale were always gauged at 282 cubical inches for the gallon, and other exciseable liquors by the wine gallon of 231 inches. They addressed a memorial to the Lords of the Treasury, stating these facts, and that, being informed the true standard wine gallon ought to contain only 224 cubical inches, they had applied to the Auditor and Chamberlains of the *Exchequer*, who, upon examination of the standard measures in their custody, had found three standard gallons, one of Henry the Seventh, and two of 1601, which an able artist employed by them had found to contain each 272 cubical inches; that, finding no wine gallon at the Exchequer, they had applied to the Guildhall of the city of London, where they were informed the true standard of the wine gallon was, and they had found, by the said artist, that the same contained 224 cubical inches only; and they added, that the gallons of other parts of the kingdom used for wine, had been made and taken from the Guildhall gallon.

In consequence of this memorial, the Lords of the Treasury, the 21st May, 1688, directed an authority to be drawn for gauging according to the Guildhall gallon; which was accordingly done; but the authority does not appear to have been signed. The ale gallon at the Treasury was of 282 inches; but the order of the Lords of the Treasury for the benefit of the revenue would have reduced the gallon, both for malt liquors and wine, to the Guildhall gallon of 224. The merchants immediately took the alarm, and petitioned that they might be allowed to *sell* by the same gauge, of 224 inches to the gallon, by which they were to be required to pay the customs and excise. The commissioners of the customs not agreeing with those of the excise, on the proposal for a new gauge, the opinion of Sir Thomas Powis, the attorney-general, was taken upon it, who advised against any departure from the usage of gauging, because the Guildhall gallon was not recognized as a legal standard, and because by any of those at the Exchequer the king would be *vastly a loser.*

Sir Thomas Powis then refers to the old statutes, prescribing that eight pounds should make a gallon; and particularly to that of 1496, requiring that the eight pounds should be of wheat; *and as* there was to be *one measure throughout the kingdom,* which could not be, unless it was adjusted by some one thing, and that seemed to be intended wheat, therefore *he did not know* how 231 cubical inches came to be taken up, but did not think it safe to depart from *the usage,* and therefore the proposal was dropped.

Sir Thomas Powis's *reasoning,* upon the statute of 1496, was perfectly correct. That statute, as well as many others, does ordain one measure throughout the kingdom; it does ordain that *every gallon* shall contain eight pounds troy weight of wheat of thirty-two kernels to the pennyweight troy, which it strangely calls a sterling. Sir Thomas did not know how 231 inches came to be taken up, because he did not know that the statute of 1496 had substituted the troy for the old easterling weight in the composition of the gallon. It was that change that brought up the 231 inches: for, if eight easterling twelve-ounce pounds of wheat filled a gallon of 217.6 inches, eight troy pounds of the same wheat must of necessity fill a gallon of 231.

The Guildhall wine gallon contained also eight troy pounds of wheat; but it was wheat thirty-two kernels of which weighed a pennyweight troy. Every kernel on the average was $\frac{1}{15}$ heavier than that which had been used for the composition of the gallon and bushel of 1266. The average kernel being specifically heavier, a pound weight of it occupied less space: on the other hand, the corn of lighter kernel would require a greater number of kernels to make up the same weight. The gallon of 1496 was to contain 61,440 kernels, weighing in the aggregate eight pounds troy: and they would fill a space of 224 cubic inches. To make the same weight, eight pounds troy would take 65,280 kernels of the wheat of 1266: but these 65,280 kernels would fill a space of 231 cubic inches. The difference between the two was a compound of the increase of numbers and the diminution of weight.

The advice of Sir Thomas Powis was, however, followed without further inquiry, and the use of the gauging-rods was continued. But in 1700 the same inconsistency of the statutes, which in the reign of Elizabeth had bred the quarrel between the informers and the herring barrels, generated a lawsuit between commerce and revenue. It has been seen, that, by a statute of 2 Henry VI., ch. 11, confirmed by subsequent acts of 1483, 1 Richard III., ch. 13, and of 1536, 28 Henry VIII., ch. 14, it had been ordained that every *butt* or pipe of wine imported should contain 126 gallons. The original statutes had reference to the Gascoign or Bordeaux wines, the casks of which were proportioned to the ton of thirty-two cubic feet. When afterward the importation of Spanish wines became frequent, they were brought in casks of different dimensions from the assize: and the statute of Richard the Third, reciting that their butts had theretofore often been

of 140 or 132 gallons, and complaining that they had been
of late fraudulently reduced to 120 gallons or less, prescribed that they should thenceforth be of at least 126
gallons. The old fashion, of 140 gallons or more to the
butt of Malmsey and other Spanish wines, was then restored: and as the law was satisfied if the butts were of
126 gallons or more, their size beyond the usual dimensions of the Gascoign standard remained unnoticed till
the fiscal officer became interested in their contents.
When customs and excise came to call for their share of
the Malmsey, the merchants for some years paid upon the
butt as if it had contained only the 126 gallons required
by the law. But this calculation could not long suit the
revenue. An action was brought by the officers of the
customs against Mr. Thomas Barker, an importing merchant, for the duties upon sixty butts of Alicant wine, for
which he had paid, as if containing 126 gallons; but
which, in fact, contained 150 gallons each. The crown
officers showed that the butt was to contain by law 126
gallons; and Mr. Leader, the city gauger, Mr. Flamstead,
and other skilful gaugers, all agreed that a wine gallon
ought to contain 231 cubical inches, and no more; that
there was such a gallon, kept from time out of mind at
Guildhall (they were in this mistaken, for it contained
only 224 inches), that the wine gallon was less than the
corn gallon, which was of 272, and the ale gallon, which
was of 282 cubical inches.

The defendant insisted that the laws had directed that
a standard should be kept at the Treasury; that there was
one there, containing 282 cubic inches; that by that measure he had paid the duty; that the Guildhall gallon was

no legal standard: and merchants, masters of ships, and vintners, of twenty, thirty, forty years' experience, all testified that Spanish wine always came in butts of 140 or 150 gallons or more. Whether Mr. Thomas Barker, when he came to *sell* his wine, retained his contempt for the Guildhall gallon, is not upon the record.

After a trial of five hours, the attorney-general made it a drawn battle; agreed to withdraw a juror; and advised to leave the remedy to parliament: and this was the immediate occasion of the statute of 5 Anne, ch. 27, sec. 17, by which the capacity of the wine gallon is fixed, and has ever since remained, at 231 cubical inches. This act declares, that any round vessel, commonly called a cylinder, having an even bottom, and being seven inches diameter throughout, and six inches deep from the top of the inside to the bottom, or any vessel containing 231 cubical inches and no more, shall be deemed and taken to be a lawful wine gallon: and it is hereby declared, that 252 gallons, consisting each of 231 cubical inches, shall be deemed a tun of wine, and that 126 such gallons shall be deemed a butt or pipe of wine, and that 63 such gallons shall be deemed an hogshead of wine.

By an act of 13 William III., ch. 5, in 1701, the Winchester bushel had been declared the standard for the measure of grain; and any cylindrical vessel of $18\frac{1}{2}$ inches diameter and 8 inches deep, was made a legal bushel. By a subsequent statute of 12 Anne, ch. 17, sec. 11, the bushel for measuring coal was to be of $19\frac{1}{2}$ inches diameter from outside to outside, and was to contain a quart of water more than the Winchester bushel; which made it of 2217.62 cubical inches.

There are several late acts of parliament (1805, 45 George III.) which mention $272\frac{1}{4}$ cubic inches as the contents of the Winchester gallon, making a bushel of 2,178 inches; and others which recognize the existence of measures different from any of the legal standards of the exchequer. By an act of 31 George III., ch. 3, inspectors of corn returns are to make a comparison between the Winchester bushel and the measure commonly used in the city or town of their inspection, and to cause a statement in writing of such comparison to be hung up in some conspicuous place.

By these successive statutes, determining in cubic inches the capacity of the vessels by which certain specific articles shall be measured, the measures bearing the same denomination, but of different contents, are multiplied; and every remnant of the original uniformity of proportion has disappeared, with the exception of that between the wine and ale gallons, and that between the troy and avoirdupois weights.

By the English system of weights and measures before the statute of 1496, the London quarter of a ton was the one measure, to which the bushel for corn, the gallon, deduced by measure, for ale, and the gallon, deduced by weight, for wine, were all referred. The hogshead was a vessel deduced from the cubing of linear measure, containing sixty-three gallons, and measuring eight cubic feet. The gallon thus formed, contained 219.43 cubic inches. This wine gallon, by another law, was to contain eight twelve-ounce pounds of wheat. One such pound of wheat, therefore, occupied 27.45 cubic inches. The vessel of eight times 27.45 cubic inches filled with wine, the liquor would weigh 54,857.1 grains of troy weight: and the weight of eight such gallons of wine would be 438,856.8 grains troy. The specific gravity of wine being to that of wheat as 175 to 143, the bushel thus formed would be of 2148.5 cubic inches; and its eighth part, or ale gallon, would be of 268.5 inches. This is only two inches more than the standard Winchester bushel of the exchequer was found to contain, and two inches less than the bushel as prescribed by the act of 13 William III.; a difference which a variation in the temperature of the atmosphere is of itself adequate to produce. It proves, that the Winchester bushel has not without reason been preserved as the favorite of all standards, in spite of all the changes, errors, and inconsistencies of legislation. But it also proves, that the ale and corn gallon ought to have continued as they originally were, of 268½ inches, and the wine gallon 219½.

The troy and avoirdupois weights are in the proportions to each other of the specific gravity of wheat and of springwater. The twelve and fifteen ounce easterling pounds were intended to be proportional between the gravity of wheat and wine. But they were roughly settled proportions, estimating the weight of wheat to be to that of wine as four to five, and the gravity of wine and of water to be the same. Under the statute of 1496, the wine gallon was of 224 inches. If troy weight was to be introduced, a gallon of this capacity had the great advantage upon which the proportion of uniformity had originally been established. The gallon contained exactly eight pounds avoirdupois of wine. The pint of wine, was a pound of wine. The corn gallon of 272 inches, corresponding with it, had the same advantage. It was filled with eight pounds of corn: a pint of wheat was a pound of wheat; and the bushel of 2,176 inches contained 64 pounds avoirdupois of that wheat, 32 kernels of which weighed one pennyweight troy. But the hogshead being of eight cubic feet, could have contained only 61¾ gallons, and the ton would have been of 247.

The wine and ale gallons, now established by law, of 231 and 282 inches, are still in the same proportion to each other as the troy and avoirdupois weights: but neither of them is in any useful proportion to the bushel. The corn gallon only is in proportion to the bushel. Neither the wine nor the corn gallons are in any useful proportion either to the weights or the coins. But the troy and avoirdupois weights are, with all the exactness that can be desired, standards for each other: and the cubic foot of spring-water weighs exactly 1000 ounces avoirdupois, by which means the ton, of thirty-two cubic feet measure, is in weight exactly 2,000 pounds avoirdupois.

Such was originally the system of English weights and measures, and such is it now in its ruins. The substitution of cubic inches, to settle the dimensions of the gallons and bushels, which began with the last century, was a change of the *test* of their contents from gravity to extension. They had before been measured by number, weight, and measure: they are now measured by measure alone. This change has been of little use in promoting the principle of uniformity. As it respects the natural standard, it has only been a change from the weight of a kernel of wheat to the length of a kernel of barley: and although it has specified the particular standard bushels and gallons, selected among the variety, which the inconsistencies of former legislation had produced, it has very unnecessarily brought in a third gallon measure quite incompatible with the primitive system ; and it has legalized two bushels of different capacity, so slightly different as to afford every facility to the fraudulent substitution of the one for the other; yet, in the measurement of quantities, resulting in a difference of between three and four per cent.

No further change in this portion of English legislation has yet been made. But the philosophers and legislators of Britain have never ceased to be occupied upon weights and measures, nor to be stimulated by the passion for uniformity. In speculating upon the theory, and in making experiments upon the existing standards of their weights and measures, they seem to have considered the principle of *uniformity* as exclusively applicable to identity, and to have overlooked or disregarded the uniformity of proportion. They found a great variety of standards differing from each other: and instead of searching for the causes of these varieties in the errors and mutability of the law, they ascribed them to the want of an immutable standard from *nature.* They felt the convenience and the facility of decimal arithmetic for *calculation ;* and they thought it susceptible of equal application to the divisions and multiplications of *time, space,* and *matter.* They despised the primitive standards assumed from the stature and proportions of the human body. They rejected the secondary standards, taken from the productions of nature most essential to the subsistence of man ; the articles for ascertaining the quantities of which, weights and measures were first found necessary. They tasked their ingenuity and their learning to find, in matter or in motion, some *immutable* standard of linear measure, which might be assumed as the single universal standard from which all measures and all weights might be derived. In the review of the proceedings in France relative to this subject, we shall trace the progress and note the results hitherto of these opinions, which have there been embodied into a great and beautiful system. In England they have been indulged with more caution, and more regard to the preservation of existing things.

From the year 1757 to 1764, in the years 1789 and 1790, and from the year 1814 to the present time, the British parliament have, at three successive periods, instituted inquiries into the condition of their own weights and measures, with a view to the reformation of the system, and to the introduction and establishment of greater uniformity. These inquiries have been pursued with ardor and perseverance, assisted by the skill of their most eminent artists, by the learning of their most distinguished philosophers, and by the contemporaneous admirable exertions, in the same cause of uniformity, of their neighboring and rival nation.

Nor have the people, or the Congress of the United States, been regardless of the subject, since our separation from the British empire. In their first confederation, these associated States, and in their present national constitution, the people, that is, on the only two occasions upon which the collective voice of this whole Union, in its constituent character, has spoken, the power of *fixing* the standard of weights and measures throughout the United

States has been committed to Congress. A report, worthy of the illustrious citizen by whom it was prepared, and, embracing the principles most essential to uniformity, was presented in obedience to a call from the House of Representatives of the first Congress of the United States. The eminent person who last presided over the Union, in the parting message by which he announced his intention of retiring from public life, recalled the subject to the attention of Congress with a renewed recommendation to the principle of decimal divisions. Elaborate reports, one from a committee of the Senate in 1793, and another from a committee of the House of Representatives, at a recent period, have since contributed to shed further light upon the subject: and the call of both Houses, to which this report is the tardy, and yet too early answer, has manifested a solicitude for the improvement of the existing system, equally earnest and persevering with that of the British parliament, though not marked with the bold and magnificent characters of the concurrent labors of France.

After a succession of more than sixty years of inquiries and experiments, the British parliament have not yet acted in the form of law. After nearly forty of the same years of separate pursuit of the same object, *uniformity*, the Congress of the United States have shown the same cautious deliberation: they have yet authorized no change of the existing law. That neither country has yet changed its law, is, perhaps, a fortunate circumstance, in reference to the principle of uniformity, for both. If this report were authorized to speak to both nations, as it is required to speak to the legislature of one of them, on a subject in which the object of pursuit is the same for both, and the interest in it common to both, it would say—Is your object *uniformity?* Then, before you change any part of your system, such as it is, compare the uniformity that you must lose, with the uniformity that you may gain, by the alteration. At this hour, fifteen millions of Britons, who, in the next generation, may be twenty, and ten millions of Americans, who, in less time, will be as many, have the same legal system of weights and measures. Their mile, acre, yard, foot, and inch—their bushel of wheat, their gallon of beer, and their gallon of wine, their pound avoirdupois, and their pound troy, their cord of wood, and their ton of shipping, are the same. They are of the nations of the earth, the two, who have with each other the most of that intercourse which requires the constant use of weights and measures. Any change whatever in the system of the one, which would not be adopted by the other, would destroy all this existing uniformity. Precious, indeed, must be that uniformity, the mere promise of which, obtained by an alteration of the law, would more than compensate for the abandonment of this.

If these ideas should be deemed too cold and cheerless for the spirit of theoretical improvement; if Congress should deem their powers competent, and their duties imperative, to establish uniformity as respects weights and measures in its most universal and comprehensive sense; another system is already made to their hands. If that universal uniformity, so desirable to human contemplation, be an obtainable perfection, it is now attainable *only* by the adoption of the new French system of metrology, in all its important parts. Were it even possible to construct another system, on different principles, but embracing in equal degree all the great elements of uniformity, it would still be a system of diversity with regard to France, and all the followers of her system. And as she could not be expected to abandon that, which she has established at so much expense, and with so much difficulty, for another, possessing, if equal, not greater advantages, there would still be two rival systems, with more desperate chances for the triumph of uniformity by the recurrence to the same standard of all mankind.

The system of modern France originated with her Revolution. It is one of those attempts to improve the condition of human kind, which, should it even be destined ultimately to fail, would, in its failure, deserve little less admiration than in its success. It is founded upon the following principles:

1. That all weights and measures should be reduced to one *uniform* standard of linear measure.

2. That this standard should be an aliquot part of the circumference of the globe.

3. That the unit of linear measure, applied to matter, in its three modes of extension, length, breadth, and thickness, should be the standard of all measures of length, surface, and solidity.

4. That the cubic contents of the linear measure, in distilled water, at the temperature of its greatest contraction, should furnish at once the standard weight and measure of capacity.

5. That for everything susceptible of being measured or weighed, there should be only one measure of length, one weight, one measure of contents, with

their multiples and subdivisions exclusively in decimal proportions.

6. That the principle of decimal division, and a proportion to the linear standard, should be annexed to the coins of gold, silver, and copper, to the moneys of account, to the division of *time*, to the barometer and thermometer, to the plummet and log lines of the sea, to the geography of the earth and the astronomy of the skies; and, finally, to everything in human existence susceptible of comparative estimation by weight or measure.

7. That the whole system should be equally suitable to the use of all mankind.

8. That every weight and every measure should be designated by an appropriate, significant, characteristic name, applied exclusively to itself.

This system approaches to the ideal perfection of *uniformity* applied to weights and measures; and, whether destined to succeed, or doomed to fail, will shed unfading glory upon the age in which it was conceived, and upon the nation by which its execution was attempted, and has been in part achieved. In the progress of its establishment there, it has been often brought in conflict with the laws of physical and of moral nature; with the impenetrability of matter, and with the habits, passions, prejudices, and necessities of man. It has undergone various important modifications. It must undoubtedly still submit to others, before it can look for universal adoption. But if man upon earth be an improveable being; if that universal peace, which was the object of a Saviour's mission, which is the desire of the philosopher, the longing of the philanthropist, the trembling hope of the Christian, is a blessing to which the futurity of mortal man has a claim of more than mortal promise; if the Spirit of Evil is, before the final consummation of things, to be cast down from his dominion over men, and bound in the chains of a thousand years, the foretaste here of man's eternal felicity; then this system of common instruments, to accomplish all the changes of social and friendly commerce, will furnish the links of sympathy between the inhabitants of the most distant regions; the metre will surround the globe in use as well as in multiplied extension; and one language of weights and measures will be spoken from the equator to the poles.

The establishment of this system of metrology forms an era, not only in the history of weights and measures, but in that of human science. Every step of its progress is interesting: and as a statement of all the regulations in France concerning it is strictly within the scope of the requisitions of both Houses, a rapid review of its origin, progress, and present state, with due notice of the obstacles which it has encountered, the changes through which it has passed, and its present condition, is deemed necessary to the performance of the duty required by the call.

In the year 1790, the present Prince de Talleyrand, then Bishop of Autun, distributed among the members of the constituent assembly of France a proposal, founded upon the excessive diversity and confusion of the weights and measures then prevailing all over that country, for the reformation of the system, or rather, for the foundation of a new one upon the principle of a single and universal standard. After referring to the two objects which had previously been suggested by Huyghens and Picard, the pendulum and the proportional part of the circumference of the earth, he concluded by giving the preference to the former, and presented the project of a decree. First, that exact copies of all the different weights and elementary measures, *used* in every town of France, should be obtained and sent to Paris: Secondly, that the national assembly should write a letter to the British parliament, requesting their concurrence with France in the adoption of a natural standard for weights and measures, for which purpose Commissioners, in equal numbers from the French Academy of Sciences and the British Royal Society, chosen by those learned bodies, respectively, should meet at the most suitable place, and ascertain the length of the pendulum at the 45th degree of latitude, and from it an invariable standard for all measures and weights: Thirdly, that, after the accomplishment, with all due solemnity, of this operation, the French Academy of Sciences should fix with precision the tables of proportion between the new standards and the weights and measures previously used in the various parts of France; and that every town should be supplied with exact copies of the new standards, and with tables of comparison between them and those of which they were to supply the place. This decree, somewhat modified, was adopted by the assembly, and, on the 22d of August, 1790, sanctioned by Louis the Sixteenth. Instead of writing to the British parliament themselves, the assembly requested the king to write to the king of Great Britain, inviting him to propose to the parliament

the formation of a joint commission of members of the
Royal Society and of the Academy of Sciences, to ascertain the natural standard in the length of the pendulum.
Whether the forms of the British constitution, the temper
of political animosity then subsisting between the two
countries, or the convulsions and wars which soon afterward ensued, prevented the acceptance and execution of
this proposal, it is deeply to be lamented that it was not
carried into effect. Had the example once been set of a
concerted pursuit of the great common object of *uniformity*
of weights and measures, by two of the mightiest and most
enlightened nations upon earth, the prospects of ultimate
success would have been greatly multiplied. By no other
means can the uniformity, with reference to the persons
using the same system, be expected to prevail beyond the
limits of each separate nation. Perhaps when the spirit
which urges to the improvement of the social condition
of man, shall have made further progress against the
passions with which it is bound, and by which it is
trammelled, then may be the time for reviving and extending that generous and truly benevolent proposal of
the constituent national assembly of France, and to call
for a *concert* of civilized nations to establish one uniform
system of weights and measures for them all.

The idea of associating the interests and the learning
of other nations in this great effort for common improvement was not confined to the proposal for obtaining the
concurrent agency of Great Britain. Spain, Italy, the
Netherlands, Denmark, and Switzerland were actually
represented in the proceedings of the Academy of Sciences
to accomplish the purpose of the national assembly. But,

in the first instance, a committee of the Academy of Sciences, consisting of five of the ablest members of the academy and most eminent mathematicians of Europe, Borda, Lagrange, Laplace, Monge, and Condorcet, were chosen, under the decree of the assembly, to report to that body upon the selection of the natural standard and other principles proper for the accomplishment of the object. Their report to the academy was made on the 19th of March, 1791, and immediately transmitted to the national assembly, by whose orders it was printed. The committee, after examining three projects of a natural standard, the pendulum beating seconds, a quarter of the equator, and a quarter of the meridian, had, on a full deliberation, and with great accuracy of judgment, preferred the last, and proposed that its ten-millionth part should be taken as the standard unit of linear measure; that, as a second standard of comparison with it, the pendulum vibrating seconds at the 45th degree of latitude should be assumed, and that the weight of distilled water at the point of freezing, measured by a cubical vessel in decimal proportion to the linear standard, should determine the standard of weights and of vessels of capacity.

For the execution of this plan they proposed six distinct scientific operations to be performed by as many separate committees of the academy.

1. To measure an arc of the meridian from Dunkirk to Barcelona, being between nine and ten degrees of latitude, including the 45th, with about six to the north and three to the south of it, and to make upon this line all suitable astronomical observations.

2. To measure anew the bases which had served before

for the admeasurement of a degree in the construction of the map of France.

3. To verify by new observations the series of triangles which had been used on the former occasion, and to continue to Barcelona.

4. To make, at the 45th degree of latitude, at the level of the sea, in vacuo, at the temperature of melting ice, observations to ascertain the number of vibrations in a day of a pendulum equal to the ten-millionth part of the arc of the meridian.

5. To ascertain, by new experiments, carefully made, the weight in vacuo of a given mass of distilled water at the freezing-point.

6. To form a scale and tables of equalization between the new measures and weights proposed, and those which had been in common use before.

This report was sanctioned by a decree of the assembly, and four committees of the academy were appointed: the three first of those enumerated objects having been intrusted to one committee, consisting of Mechain and Delambre. The experiments upon the pendulum were committed to Borda, Mechain, and Cassini; those on the weight of water to Lefevre Gineau, and Fabbroni; and the scale and tables to a large committee on weights and measures.

The performance of all these operations was the work of more than seven years. Two of them, the measurement of the arc of the meridian, and the ascertainment of the specific gravity of water in vacuo, were works requiring that combination of profound learning which is possessed of the facts in the recondite history of nature already ascertained, with that keenness of observation which detects facts still deeper hidden ; that fertility of genius which suggests new expedients of invention, and that accuracy of judgment which turns to the account, not only of the object immediately sought, but of the general interests of science, every new fact observed. The actual admeasurement of an arc of the meridian of that extent had never before been attempted. The weighing of distilled water in vacuo had never before been effected with equal accuracy. And in the execution of each of these works, nature, as if grateful to those exalted spirits who were devoting the labors of their lives to the knowledge of her laws, not only yielded to them the object which they sought, but disclosed to each of them another of her secrets. She had already communicated by her own inspiration to the mind of Newton, that the earth was not a perfect sphere, but an oblate spheroid, flattened at the poles ; and she had authen- . ticated this discovery by the result of previous admeasurements of degrees of the meridian in different parts of the two hemispheres. But the proportions of this flattening, or, in other words, the difference between the circles of the meridian and equator, and between their respective diameters, had been variously conjectured from facts previously known. To ascertain it with greater accuracy was one of the tasks assigned to Delambre and Mechain ; for, as it affected the definite extension of the meridian circle, the length of the metre, or aliquot part of that circle which was to be the standard unit of weights and measures, was also proportionably affected by it. The result of the new admeasurement was to show that the flattening was of $\frac{1}{334}$; or that the axis of the earth was to the diameter of the equator as 333 to 334. Is this proportion to the decimal

number of 1000 accidental? It is confirmed as matter of
fact, by the existing theories of astronomical nutation and
precession, as well as by experimental results of the length
of the pendulum in various latitudes. Is it also an index
to another combination of extension, specific gravity, and
numbers, hitherto undiscovered? However this may be,
the fact of the proportion was, on this occasion, the only
object sought. This fact was attested by the diminution
of each degree of latitude, in the movement from the north
to the equator; but the same testimony revealed the new
and unexpected fact, that the diminution was not regular
and gradual, but very considerably different at different
stages of the progress in the same direction; from which
the inference seems conclusive, that the earth is no more
in its breadth than in its length, perfectly spherical, and
that the northern and southern hemispheres are not of
dimensions precisely equal.

The other discovery was not less remarkable. The
object to be ascertained was the specific gravity of a given
mass of water in vacuo, and at its *maximum* of density;
that is, at the temperature where it weighs most in the
smallest space. That fluids are subject to the general laws
of expansion and contraction from heat and cold, was the
principle upon which the experiments were commenced.
It was also known that, in the transition of fluids to a
solid state, the reverse of this phenomenon occurs, and
that water, in turning to ice, instead of contracting, expands. It had been supposed that the freezing-point was
that at which this polarity of heat and cold, if it may be
so called, was inverted, and that water, contracting as it
cooled until then, began at once to freeze and to expand.

The discovery made by Lefevre Gineau, and Fabbroni, was, that the change took place at an earlier period; that water contracts as it cools, till at five degrees above 0 of the centigrade, answering to forty-one of Fahrenheit's thermometer, and, from that term, gradually expands as it grows cold, till fixed in ice at 0 of the former, or thirty-two degrees of the latter.

In the admeasurement of the arc of the meridian, and in the weighing of the given volume of water, the standard measure and weights, previously established by the laws of France, were necessarily used. The identical measure was a toise or fathom belonging to the Academy of Sciences, which had been used for the admeasurement of several degrees of the meridian between the years 1737 and 1741, in Peru, and had thence acquired the denomination of the *Toise du Perou*. In 1766 it had served as the standard from which eighty others had been copied, and sent to the principal bailiwicks in France, and to the châtelet at Paris. The instruments used by Delambre and Mechain, for their mensurations, were two platina rods, each of double the length of this fathom of Peru. A repeating circle, a levelling instrument, and a metallic thermometer, consisting of two blades, one of brass, and the other of platina, and calculated to show the difference of expansion produced upon the two metals by the ordinary alternations of heat and cold in the atmosphere, all invented by the ingenious and skilful artist Borda, were also among the instruments used by the commissioners.

The weights with which the new standard was compared, were a pile of fifty marks, or twenty-five Paris pounds, called the weights of Charlemagne, and which, though not of the antiquity of that prince's age, had been used as standards for a period of more than five hundred years.

The fathom of Peru was divided into six standard royal feet of France, each foot into twelve thumbs, each thumb into twelve lines. The toise, therefore, was of seventy-two thumbs, or 864 lines. The standard metre of platina, the ten-millionth part of the quarter of the meridian, measured by the brass fathom of Peru, was found to be equal to 443 lines, and 295,936 decimal parts of a line: and as it was found impossible to fix in the concrete form a division smaller than the thousandth part of a line, the definitive length of the metre was fixed at 443,296 lines, equivalent, by subsequent experiments of the academy, to 39.3827 English inches; by the latest experiments of Captain Kater to 39.37079; and by those of Mr. Hassler, in this country, to 39.3802.

The Paris pound, mark-weight as it was called (poids de marc), of the pile of Charlemagne, consisted of two marks, each mark of eight ounces, each ounce of eight gros or drams, each gros of three deniers or pennyweights, and each denier of twenty-four grains. The pound, therefore, consisted of 9,216 grains, and was equal to fifteen ounces and fifteen pennyweights, or 7,560 grains troy. The grain was rather more than four-fifths of the troy grain, and had probably, in the origin, been equivalent to the kernel of wheat, which the troy grain could scarcely have been. The cubic decimetre, or tenth part of the metre, of distilled water, at the temperature of its greatest density, weighed in vacuo, was found of equal weight with 18.827 grains $\frac{15}{100}$ of a grain; or two pounds, five gros,

thirty-five grains $\frac{15}{100}$ of the mark-weight: and this, by
the name of the kilogramme, was made the standard
weight, its thousandth part being the gramme, or unit,
equivalent to 15.44572 grains troy, or about two and one-
fifth pounds avoirdupois.

The capacity of the vessel containing this water was at
the same time made the standard of all measures, liquid
or dry: it was called a *litre*, and is of the contents of
61.0271 cubic inches, about one-twentieth more than our
wine quart. The metre was applied to superficial and solid
measures, according to their proportions: the chain of ten
metres being applied to land measure, and its square
denominated an *are ;* the cubic metre was called a *stere.*

The principle of decimal arithmetic was applied exclusively to all these weights and measures: their multiples
were all tenfold, and their subdivisions were all tenth
parts.

To complete the system, a vocabulary of new denominations was annexed to every weight and measure belonging to it. As a circumstance of great importance to the
final success of the system, it may be remarked that these
two incidents, the exclusive adoption of decimal divisions,
and the new nomenclature, have proved the greatest obstacles to the general introduction of the new weights and
measures among the people.

It has indeed from its origin, like all great undertak-
ings, been obliged to contend with the intemperate zeal
and precipitation of its friends, not less than with prejudice, ignorance, and jealousy, of every description. The
admeasurement of the meridian was commenced at the
very moment of the fanatical paroxysm of the French

revolution. At every station of their progress in the field survey, the commissioners were arrested by the suspicions and alarms of the people, who took them for spies, or engineers of the invading enemies of France. The government was soon overthrown; the Academy of Sciences abolished; and the national assembly of the first constitutional monarchy, just at the eve of their dissolution, instead of waiting calmly for the completion of the great work which was to lay the foundation for a system to be as lasting as the globe, in a fit of impatience passed, on the 1st of August, 1793, a law declaring that the system should go immediately into operation, and assuming for the length of the standard metre the ten-millionth part of the quadrant of the meridian according to the result of the old admeasurement of a degree in 1740, and arranging an entire system of weights and measures, in decimal divisions, with new denominations, all of which were to be merely temporary, and to cease when the definitive length of the metre should be ascertained. This extraordinary act was probably intended, as it directly tended, to prevent the further prosecution of the original plan: and though, soon after, it was followed by a decree of 11th September, 1793, authorizing the temporary continuance of the general committee of weights and measures, which had been appointed by the academy, yet, on the 23d December of the same year, a decree of Robespierre's committee of public safety dismissed from the commission Borda, Lavoisier, Laplace, Coulomb, Brisson, and Delambre, on the pretence that they were not republicans sufficiently pure. Mechain escaped the same proscription only because he was detained as a prisoner in Spain.

Yet even Robespierre and his committee were ambitious, not only of establishing the system of new weights and measures in France, but of offering them to the adoption of other nations. By a decree of that committee of 11th December, 1793, the board or commission of weights and measures were directed to send to the United States of America a metre in copper and a weight, being copies of the standards then just adopted. They were accordingly transmitted: and on the 2d of August, 1794, the two standards were, by the then French minister plenipotentiary Fauchet, sent to the Secretary of State, with a letter, recommending, with some urgency, the adoption of the system by the United States. This letter was communicated to Congress by a message from the President of the United States, of the 8th of January, 1795.

In the mean time the mensuration of the arc of the meridian was entirely suspended by the dismissal of Delambre, and the detention of Mechain. Its progress was renewed by a decree of the national convention of 7th April, 1795 (18 Germinal, An. 3), which abolished almost entirely the nomenclature of the temporary standards adopted in August, 1793, and substituted a new one, being that still recognized by the law, and the units of which have been already mentioned; the metre, the gramme, the are, the litre, and the stere. To express the multiples of these units, the Greek words denominating ten, a hundred, a thousand, and ten thousand, were prefixed as additional syllables, while their tenth, hundredth, and thousandth parts were denoted by similar prefixed syllables from the Latin language. Thus, the myria-metre is ten thousand, and the kilo-metre one thousand, the hecto-metre one hundred, and the deca-metre ten metres; each of those prefixed syllables being the Greek word expressive of those respective numbers; while the deci-metre, the centi-metre, and the milli-metre, are tenth, hundredth, and thousandth parts, signified by the Latin syllables respectively prefixed to them. The theory of this nomenclature is perfectly simple and beautiful. Twelve new words, five of which denote the things, and seven the numbers, include the whole system of metrology; give distinct and significant names to every weight, measure, multiple, and subdivision of the whole system; discard the worst of all the sources of error and confusion in weights and measures, the application of the same name to different things; and keep constantly present to the mind the principle of decimal arithmetic, which combines all the weights and measures, the proportion of each weight or measure with all its multiples and divisions, and the chain of uniformity which connects together the profoundest researches of science with the most accomplished labors of art and the daily occupations and wants of domestic life in all classes and conditions of society. Yet this is the part of the system which has encountered the most insuperable obstacles in France. The French nation have refused to learn, or repeat these twelve words. They have been willing to take a total and radical change of things; but they insist upon calling them by old names. They take the metre; but they must call one-third part of it a foot. They accept the kilogramme; but, instead of pronouncing its name, they choose to call one-half of it a pound. Not that the third of a metre is a foot, or the half of a kilogramme is a pound· but because they are not very different from them, and because, in expressions of popular origin, distinctness of idea in the use of language is more closely connected with habitual usage than with precision of expression.

This observation may be illustrated by our own experience, in a change effected by ourselves in the denominations of our coins, a revolution by all experience known to be infinitely more easy to accomplish than that of weights and measures. At the close of our war for independence, we found ourselves with four English words, pound, shilling, penny, and farthing, to signify all our moneys of account. But, though English words, they were not English things. They were nowhere sterling: and scarcely in any two States of the Union were they representatives of the same sums. It was a Babel of confusion by the use of four words. In our new system of coinage we set them aside. We took the Spanish piece of eight, which had always been the coin most current among us, and to which we had given a name of our own—a dollar. Introducing the principle of decimal divisions, we said, a tenth part of our dollar shall be called a *dime*, a hundredth part a *cent*, and a thousandth part a *mille*. Like the French, we took all these new denominations from the Latin language; but instead of prefixing them as syllables to the generic term dollar, we reduced them to monosyllables, and made each of them significant by itself, without reference to the unit of which they were fractional parts. The French themselves, in the application of their system to their coins, have followed our example; and, assuming the franc for their unit, call its tenth part a *decime*, and its hundredth a *centime*. It is now nearly thirty years since our new moneys of account, our coins, and our mint, have been established. The dollar, under its new stamp, has preserved its name and circulation. The cent has become tolerably familiarized to the tongue, wherever it has been made by circulation familiar to the hand. But the dime having been seldom, and the mille never, presented in their material images to the people, have remained so utterly unknown, that now, when the recent coinage of dimes is alluded to in our public journals, if their name is mentioned, it is always with an explanatory definition to inform the reader, that they are ten-cent pieces; and some of them which have found their way over the mountains, by the generous hospitality of the country, have been received for more than they were worth, and have passed for an eighth, instead of a tenth, part of a dollar. Even now, at the end of thirty years, ask a tradesman, or shopkeeper, in any of our cities what is a dime or a mille, and the chances are four in five that he will not understand your question. But go to New York and offer in payment the Spanish coin, the unit of the Spanish piece of eight, and the shop or market-man will take it for a *shilling.* Carry it to Boston or Richmond, and you shall be told it is not a shilling, but nine pence. Bring it to Philadelphia, Baltimore, or the City of Washington, and you shall find it recognized for an eleven-penny bit; and if you ask how that can be, you shall learn that, the dollar being of ninety pence, the eighth part of it is nearer to eleven than to any other number: and pursuing still further the arithmetic of popular denominations, you will find that half eleven is five, or, at least, that half the eleven-penny bit is the fi-penny bit, which fi-penny bit at Richmond shrinks to four pence half-penny, and at New York swells to six pence. And thus we have English denominations most absurdly and diversely applied to Spanish coins; while our own lawfully established dime and mille remain, to the great mass of the people, among the hidden mysteries of political economy—state secrets.

Human nature, in its broadest features, is everywhere the same. This result of our own experience, upon a small scale, and upon a single object, will easily account for the repugnance of the French people to adopt the new nomenclature of their weights and measures. It is not the length of the words that constitutes the objection against them, nor the difficulty of pronunciation; for, fi-penny bit is as hard to speak and as long a word as kilogramme, and eleven-penny bit has certainly more letters and syllables, and less euphony, than myria-metre. But it is because, in the ordinary operations of the mind, distinctness of idea is, by the laws of nature, linked with the chain of association between sensible images and their habitual denominations, more closely than with the exactness of logical analysis.

The nomenclature of the French metrology was established by the law of 7th April, 1795, although the metre and the kilogramme were only provisional and not definitive. It was known that the difference between the provisional and definitive metre and kilogramme would be very small, scarcely perceptible: and by that inverted logic which presides over all precipitate legislation, it was concluded that because it was small it would be unimportant: instead of which, sound reason would have inferred, that, to the purpose of uniformity, the smaller the difference was, the greater was the danger of its producing confusion between the temporary and the perpetual things which were to bear the same name.

But with the hasty call for a provisional metre and kilogramme, the law of 7th April, 1795, gave the definitive nomenclature, and directed the renewal of all the operations commenced under the direction of the Academy of Sciences; and the persons employed upon them were reinstated in their functions by the committee of public instruction of the national convention. A commission of twelve persons, Berthollet, Borda, Brisson, Coulomb, Delambre, Haüy, Lagrange, Laplace, Mechain, Monge, Prony, and Vandermonde, was appointed on the 17th of April, 1795, for the final accomplishment of the original plan; the most important and laborious part of which, the admeasurement of the arc of the meridian, was immediately resumed by Delambre and Mechain. By them the whole distance from Dunkirk to Mont Jouy, near Barcelona, a distance of nine degrees and two-thirds, more than a tenth part of the quadrant of the meridian, was measured by trigonometrical survey. The angles formed by every station with those next before and after it, rectified by the angles of elevation and depression formed by the inequalities on the surface of the ground, to reduce the whole to the level of the horizon, were measured and referred to the measure of two bases, one between Melun and Lieusaint, the other between Vernet and Salces, near Perpignan; each serving as a corrective upon the other. Observations of azimuth ascertained the direction of the sides of triangles, with reference to the meridian; and astronomical observations ascertained the celestial arc, corresponding with that which was measured upon the earth. The distance from Dunkirk to Rhodez, about 450 miles, was performed by Delambre; and that from Barcelona to Rhodez, upward of 200 miles, by Mechain. The base of Melun was of 6075.90 and that of Perpignan 6006.25 toises, each nearly seven miles: and though at the distance of near 400 miles from each other, the base of Perpignan, calculated by inference from the chain of triangles between them, differed from its actual admeasurement less than one foot. The portion of the distance allotted to Mechain was less than one-third of the whole; but, traversing the Pyrenees, and being chiefly upon the Spanish territories, was attended with more difficulties than those encountered by his associate. Mechain, in the execution of his task, had formed the project of extending the survey to the Balearic islands, which would have made the portion of the arc south of the forty-fifth degree equal to that northward of it. With a firmness and perseverance of pursuit, amidst innumerable obstacles, he had proceeded far in the execution of this supplementary plan. His triangles were already extended from Barcelona to Tortosa. His stations had been selected to Cullora. Six or seven triangles more would have carried his work to its termination in the island of Ivica. Arrested by a fever, in his first progress, and compelled then to abandon the attempt, he had resumed it after the result of the original plan had been ascertained, and the new system had been finally established by law. The hand which sets bounds to all human pursuits again and definitively met and closed his career. He died on the 20th of September, 1805, at Castellon de la Plana, in the Spanish province of Valencia.

His more fortunate associate, Delambre, has published, in three quarto volumes, under the title of the "Basis of the metrical decimal system, or measure of the arc of the meridian between Dunkirk and Barcelona," all the details and results of this admirable operation. A fourth volume yet remains to be published, which will contain the account of the actual execution since the death of Mechain, of the idea which he had conceived of extending the admeasurement to the island of Formentara, and of the additional extention of it northward to the Shetland Islands, by connecting it with the trigonometrical survey of Great Britain. This work, in passing to future ages, a monument of the philosophy, science, public spirit, and active benevolence of our own, will redeem, by the martyrdom of genius and learning in the cause of human happiness, the blood-polluted glories of contemporaneous war.

The reports of the proceedings of Delambre and Mechain, as well of their field surveys as of their astronomical observations, and all their calculations, were submitted to the inspection, scrutiny, and revision, of a committee of the mathematical and physical class of the national institute, that phœnix of science which had arisen from the ashes of the academy of sciences. The observations to ascertain the length of the pendulum, and the experiments for determining the specific gravity of distilled water at its maximum of density, were submitted to the same ordeal. Two reports upon the whole result were made to the class, one by Trallés of the Helvetic Confederation, the other by Van Swinden of the Netherlands, two of the foreign associates, who had been invited to co-operate in the labor,

and to participate in the honor of the undertaking. These
two reports, combined by Van Swinden into one, were
then reported from the class to the general meeting of the
institute, and by that body, with all suitable solemnity,
to the two branches of the national assembly of France,
on the 22d of June, 1799, together with a definitive metre
of platina, made by Lenoir, and a kilogramme of the same
metal, made by Fortin. They were introduced by an appropriate address at the bar of the two Houses, by the
presiding member of the national institute, La Place; to
which answers were returned by the respective presidents
of the two legislative chambers. On the same day, the
standard metre and kilogramme were deposited in the
hands of the keeper of the public archives; and a record
of the fact was made and signed by him, and by all the
members of the institute, foreign associates, and artists,
whose joint labors had contributed to the consummation
of this more than national undertaking.

No apology will be deemed necessary by Congress for
dwelling upon these details which signalized the establishment of the new French metrical system. The spectacle
is at once so rare and so sublime, in which the genius, the
science, the skill, and the power of great confederated
nations are seen joining hand in hand in the true spirit of
fraternal equality, arriving in concert at one destined stage
of improvement in the condition of human kind; that,
not to pause for a moment, were it even from occupations
not essentially connected with it, to enjoy the contemplation of a scene so honorable to the character and capacities
of our species, would argue a want of sensibility to appre-
ciate its worth. This scene formed an epocha in the history

of man. It was an example and an admonition to the
legislators of every nation, and of all after-times.

On the 10th of December, 1799 (19 Frimaire, 8), the
temporary metre and kilogramme, which had been ordained
by the laws of 1st August, 1793, and 7th April, 1795 (18
Germinal, 3), were abolished. That metre had been of 443
lines and $\frac{44}{100}$ of a line of the ancient foot, standarded by
the fathom of Peru. The new and definitive metre was
fixed at 443 lines $\frac{296}{1000}$. The difference between them was
about $\frac{1}{7}$ of a line, or $\frac{1}{100}$ of our inch, a difference imperceptible for all ordinary uses; but very important as a
standard variety, and immediately apparent when multiplied to the cube for the measure of capacity and the
weight. Thus the temporary kilogramme had been of
18,841 grains mark weight, while the new and definitive
kilogramme was reduced to 18,827.15 grains.

During the violent ebullitions of the most inflammatory
period of the French Revolution, it had been imagined
that in the reformation of the system of weights and measures, upon the principle of uniformity, the mensuration of
time ought to be included. But this was a different
project from that of the reformed metrology, originating
in motives less pure and ingenuous, connected with purposes interfering with religious impressions, and quite
inconsistent with one of the principal expedients of per-
petuating the identity of the new weights and measures.
The length of the pendulum beating seconds, it has been
seen, is, in the new metrical system, the test of verification
for that of the metre, in case the original platina standard
should be lost. The pendulum beating seconds vibrates
86,400 times in the solar day of 24 hours. But the fiery

spirits of the Revolution called for a reformation of the calendar, for a new constitution of the seasons, and above all, for decimal divisions. The establishment of the French republic was a new æra to the world. It had taken place on the 22d September, 1792, the day of the autumnal equinox, when the sun entered the sign of the Balance, the symbol of equality. Before it the Christian æra was to disappear. The new year was to commence with that day. The division of twelve months was to be retained; but they were all to be of three times ten or thirty days. The division of weeks of seven days, beginning with one specially devoted to the worship of the Creator, and repose, was to be abolished; but every tenth day was to be dedicated to some moral abstraction or virtue, such as liberty, equality, fraternity, patriotism, conjugal affection, filial piety, old age, and once a year to the Supreme Creator, whose existence was formerly authenticated by a decree of the national convention. After their thirty-six decads, there remained five, and in leap-year six, complementary days, to which they gave a name which can scarcely be repeated with decency; but which were to be all holidays, and in which were to be revived the Olympic games of ancient Greece. The names of the months were to be significant. The three successive months, composing each of the four seasons, were to have the same terminating syllable, which in its sound should convey to the ear its distinctive character. The first of the four was *aire*, which was supposed to indicate the solemn and majestic tranquillity of autumn; the second *ose*, a dull and heavy sound, marking the torpor and frigidity of winter; the third *al*, which had all the reviving influence and liquid harmony of spring; and the fourth *dor*, burning to the fancy with the vivid ardors of summer. To these terminating syllables, each month had an appropriate prefix. Thus, in autumn, Vendemi-aire, was the month of vintage; Brumaire, the month of fogs; and Frim-aire, the month of incipient cold. From the winter solstice to the vernal equinox, there was Niv-ose, the month of snow; Pluvi-ose, the month of rain; and Vent-ose, the month of wind. These were succeeded by the darlings of the year; Germin-al, the month of buds; Flore-al, the month of blossoms; and prairi-al, the month of blooming meads. The procession closed with the bounties and fervors of summer; Messi-dor, the month of harvests; Thermi-dor, the month of heat; and Fructi-dor, the month of fruit. The days of their decad were to be denominated by their numbers from one to ten; as Primedi, first day, Duodi, second day, and so on to Decadi, the tenth day; which was to be the day of relaxation from labor, and of meditations upon virtue. But the clashing of the new calendar with the new metrology was the division of the solar day, not into 24 hours, of 60 minutes, with 60 seconds to the minute; but into ten hours, each of 100 minutes, and each minute of 100 seconds. The pendulum of that day, therefore, would vibrate 100,000 times, and would be of quite a different length from that which was to be the test of verification to the metre.

This system has passed away, and is forgotten. This incongruous composition of profound learning and superficial frivolity, of irreligion and morality, of delicate imagination and coarse vulgarity, is dissolved. This statue, with the form of Apollo, and the face of Silenus, has crumbled into dust; but it was established by a law of the 5th of October, 1792, and for the space of twelve years it was the calendar of the French nation. Henceforth it will only be remembered as preparing future problems in chronology.

The division of the day into a hundred thousand parts had some reasons to recommend it, but was the first part of the system that was abandoned. It had been decreed as compulsory, with the new nomenclature of the calendar, on the 24th of November, 1793 (4 Frimaire, 2), but this regulation was indefinitely suspended by the law of 7th April, 1795 (18 Germinal, 3).

On the 8th of April, 1802, when a First Consul, soon to be for life, had produced some perturbation of that balance, the symbol of equality in which the sun had first shone upon the French Republic, there passed a law, retaining the *equinoctial* or republican calendar for all civil purposes, but resuming the *solstitial* or Gregorian calendar so far as to restore its week of seven days with their names, and its Sabbath of the first of them. The terms *equinoctial* and *solstitial*, in this law, applied to the new and old calendars, seem studiously selected to veil the balance of equality on one side, and the Sabbath of religion on the other.

But on the 9th of September, 1805 (22 Fructidor, 13), in the month of *fruits*, and when the sun of the French Republic had got, if not into the sign, at least deep into the constellation, of the Lion; when the legend of her coins bore on one side the name and head of Napoleon, Emperor, and on the other the name of the French Republic, a senatus consultum ordained, that, from the 11th of that dull and heavy month of snows of the 14th year, the 1st of January, 1806, should reappear, and the Gregorian calendar should be restored to use throughout the Republican Empire.

The decimal divisions, and the fanciful contexture of the equinoctial calendar, were a sort of episode to the new system of metrology. The attempt to decimate the year in its number of days was equally useless and absurd. The five successive holidays at the close of the year, just at the season of the vintage, with the institution of athletic sports, were a waste of time, and a provocation to mischievous idleness, ill compensated by the retrenchment of sixteen Sundays in the year, at the distance of a week from each other, and devoted to the exercises of piety.

The application of the metrical system to geography and astronomy was a much more rational part of the project, but has been attended with difficulties in execution hitherto insuperable. In adopting an aliquot decimal part of the quadrant of the meridian for the unit of long measures, it formed a natural division of the quadrant itself into ten parts, each of ten degrees. The degree would then have been of 100,000 metres, and the number of degrees to encircle the earth would have been four hundred. The degree, which is now of about sixty-nine English miles, would have been of about sixty-two, and the facility of all astronomical, geographical, and nautical calculations, would have been much increased. But it would have rendered useless all the tables indispensable to the navigator, astronomer, and geographer; and, if it had not produced the same effect upon all the maps and charts now in use, it would have tended to produce confusion between those of the old and those of the new system. The ancient division of the sphere, and, consequently, of the circle, into 360, and therefore into quadrants of 90 degrees, originated in the coincidence of the daily rotations of the earth in its orbit round the sun, or the apparent motion of the sun in the ecliptic, which, as near as the approximation of numbers can bring it, is of one degree every day. The division of the day into twenty-four hours, each of sixty minutes, is founded on a similar coincidence of time in the rotation of the earth round its axis, and the apparent daily revolution of the firmament round the earth resulting from it; giving for the rising or setting of each sign of the zodiac a term of two hours, and for each degree of the circle described by the earth in its rotation a term of four minutes, or fifteen degrees to the hour. The adoption of the decimal divisions for the quadrant of the meridian, and for the circle, would have disturbed all these harmonies, as well as that of the sexagesimal division of the circle by the radius; a division not perfectly exact, since the radius is not exactly the sixth part of the circumference, but which, having been found the most convenient for practice, has been established from the remotest antiquity, and, being already used by all the civilized nations of the earth, could not, by being set aside, tend to uniformity, unless the method to supply its place could be alike secure of universal adoption.

The divisions of the barometer had always been marked in inches and lines. The application to it of the decimetre, its multiples and divisions, had for observation and calculation the usual conveniences of the decimal arithmetic. The graduation of the thermometer had always been arbitrary and various in different countries. The principle of the instrument was everywhere the same, that of marking the changes of heat and cold in the atmosphere, by the expansion and contraction which they produced upon mercury or alcohol. The range of temperature between boiling and freezing water was usually taken for the term of graduation, but, by some, it was graduated downward from heat to cold, and by others upward from cold to heat. By some the range between the two terminating points was divided into 80, 100, 150, or 212 degrees. One put the freezing, and another the boiling point, at 0. Reaumur's thermometer, used in France, began with 0 for the freezing-point, and placed the boiling-point at 80. Fahrenheit's, commonly used in England and in this country, has the freezing-point at 32, and the boiling-point at 212. The centigrade thermometer, adopted by the new system, begins at the freezing-point at 0, and places the boiling-point at 100: its graduation, therefore, is decimal, and its degrees are to those of Reaumur as five to four, and to those of Fahrenheit as five to nine.

The application of the new metrology to the moneys and coins of France has been made with considerable success; not, however, with so much of the principle of uniformity as might have been expected, had it originally formed a part of the same project. But the reformation of the coins was separately pursued, as it has been with us; and, as the subject is of great complication, it naturally followed that, from the separate construction of two intricate systems, the adaptation of each to the other was less correct than it would have been, had all the combinations of both been included in the formation of one great masterpiece of machinery. It is to be regretted that, in the formation of a system of weights and measures, while such extreme importance was attached to the discovery and assumption of a national standard of long measure as the link of connection between them all, so little consideration was given to that primitive link of connection between them, which had existed in the identity of weights and of silver coins, and of which France, as well as every other nation in Europe, could still perceive the ruins in her monetary system then existing. Her livre tournois, like the pound sterling, was a degeneracy, and a much greater one, from a pound weight of silver, but it had scarcely a seventieth part of its original value. It was divided into twenty sols or shillings; and the sol was of twelve deniers or pence. It had become a mere money of account; but the ecu, or crown, was a silver coin of six livres, nearly equivalent to an ounce in weight, and there were half-crowns, and other subdivisions of it, being coins of one-fourth, one-fifth, one-eighth, and one-tenth, of the crown. There were also coins of gold, of copper, and of mixed metal called billon, in the ordinary circulations of exchange. Shortly after the adoption of the provisional or temporary metre and kilogramme, a law of 16 Vendemaire 2 (7th October, 1793), prescribed that the principal unit, both of gold and of silver coins, should be of the weight of ten *grammes*. The proportional value of gold to silver was retained as it had long before been established in France, at 15½ for one: the alloy of both coins was fixed at one-tenth; and the silver franc of that coinage would have been worth about thirty-eight cents, and the gold franc a little short of six dollars. This law was never carried into execution. It was superseded by one of 15th

August, 1794 (28 Thermidor, 3), which reduced the silver franc to five grammes; and it was not until after a law of 7 Germinal, 11 (28th March, 1803), that gold pieces of twenty and forty francs were coined at 155 of the former to the kilogramme.

In the new system, the name of *livre*, or pound, as applied to money or coins, was discarded: but the *franc* was made the unit both of coins and moneys of account. The franc was a name which had before been in common use as a synonymous denomination of the *livre*. The new franc was of intrinsic value $\frac{1}{80}$ more than the livre. The franc is decimally divided into *decimes* of $\frac{1}{10}$, centimes of $\frac{1}{100}$, and millimes of $\frac{1}{1000}$, of the unit; but the smallest copper coin in common use is of five centimes, equivalent to about one of our cents. The silver coins are of one-fourth, one-half, one and two francs, and of five francs; the gold pieces, of twenty and forty francs. The proportional value of copper to silver is of one to forty, and that of billon to silver of one to four: so that the kilogramme should weigh 5 francs of copper coin, 50 of the billon, 200 of the silver, and 3,100 of the gold coins: and the decime of billon should weigh precisely two grammes. The allowances, known by the name of remedy for errors in the weight and purity of the coins, are of $\frac{2}{100}$ upon copper, which is only for excess: those upon the weight of billon are of $\frac{14}{1000}$; upon silver $\frac{20}{1000}$ for one-quarter francs, $\frac{14}{1000}$ for one-half francs, and of $\frac{10}{1000}$ or one per cent. on one and two franc pieces, and of $\frac{6}{1000}$ for five-franc pieces. That of the gold coins is of $\frac{4}{1000}$; all, excepting the copper, allowances either for excess or deficiency. But the practice of the mint never transgresses in excess; and the deficiency is always nearly the whole allowed by law. The remedy of alloy is of $\frac{7}{1000}$, either of excess or defect, for billon; of $\frac{3}{1000}$ for silver; and of $\frac{2}{1000}$ for gold. It is said that the actual purity of the coins, both of gold and of silver, is within $\frac{1}{1000}$ less than the standard.

The conveniences of this system are,

First, The establishment of the same proportion of alloy to both for gold and silver coins, and that proportion decimal.

Secondly, The established proportions of value between gold, silver, mixed metal, and copper coins.

Thirdly, The adaptation of all the coins to the weights in such manner as to be checks upon and tests of each other. Thus the decime of billon should weigh two grammes, the franc of silver five, the two-franc piece of silver and the five-centime piece of copper each ten, and the five-franc piece fifty. The allowances of remedy disturb partially these proportions. These are practices continued in all the European mints, after the reasons upon which they were originally founded have in a great measure ceased. In the imperfection of the art, the mixture of the metals used in coining, and the striking of the coins, could not be effected with entire accuracy. There would be some variety in the mixture of metals made at different times, though in the same intended proportions, and in different pieces of coin, though struck by the same process and from the same die. But the art of coining metals has now attained a perfection, that such allowances have become, if not altogether, in a great measure unnecessary. Our laws make none for the deficiencies of weight: and they consider every deficiency of purity as an error, for which the officers of the mint shall be *excused* only in case of its being within $\frac{1}{144}$ part, or about $\frac{1}{1000}$; for if it should exceed that, they are disqualified from holding their offices. Where the penalty is so severe, it is proper that the allowance should be large; but, as obligatory duty upon the officers of the mint, an allowance of $\frac{1}{1000}$ would be amply sufficient for each single piece, and no allowance should be made upon the average.

Among the difficulties attending all innovations upon established usages relative to weights and measures, are their application to the tonnage of ships and boats, and to the form and size of casks. We have seen, in the review of the history of English weights and measures, how Henry the Seventh's change of the Rochelle for the troy pound affected the barrels of herring fishers, the hogshead of claret, and the butts of Alicant wine. The tonnage of ships, on the old established metrologies, was founded, like their weights and measures of capacity, upon a principle of combining specific gravity and occupied space. The ton of shipping was adapted both for a weight and a measure. The capacity of a ship as a measure is ascertained by its internal cubical dimensions, which, before the change of system in France, gave 42 royal cubic feet to a ton. The mode of admeasurement was, like ours, a complicated multiplication and division of length, breadth, and thickness, with given deductions and estimates, all finally divided by the standing number 94, as ours is by 95, and the quotient of which gives the number of what may be called custom-house *tons*. But the French ordinances, like our law, did not indicate by what specific measure this length, breadth, and thickness were to be taken. It was always perfectly understood here, that it is in feet, and tenths of feet ; and in France, that it was in royal feet and their tenths. Nothing can afford a more striking illustration of the construction which long established usage can give to law than this admeasurement in feet and tenth parts of a foot, differing from that used in all other cases of feet and inches, or twelfth parts, not expressly directed by law, and yet practised for these thirty years, propably without a question upon the meaning of the law. The attempt in France to apply it to the admeasurement by the metre, without changing the final common divisor, 95, signally shows how cautiously complications of weights, measures, numbers, and coins must be dealt with. The law of the 12th Nivose, 2 (1st January, 1794), directed those measures to be taken in the new metre and divisions, without changing the final divisor, 95, to produce a number of *tons*. The consequence would have been that the cubic numbers divided by 95 would have been metres and their decimal parts, instead of feet and their decimal parts; and the quotient would have reduced the tonnage to about one-third of its proper dimensions. To have produced a quotient of a number of tons, their quotient would have been 30, instead of 95. This mistake was precisely the same as that of the British parliament of 1496, when, thinking to re-enact the law of 1266, they prescribed a bushel to be made from sixty-four gallons *troy* weight of wheat of thirty-two kernels to the troy pennyweight, instead of a bushel of sixty-four pounds sterling at fifteen ounces to the pound, of wheat, thirty-two kernels of which weighed the penny sterling of Henry the Third. It was the same mistake which the Greek Church yearly repeats in celebrating Easter by the Julian calendar of 365 days 6 hours to the year, and the lunar cycle of nineteen years. And to come nearer home to ourselves, it was the same mistake which our own statute-book discloses in estimating the British pound sterling four dollars forty-four cents, because one hundred and ten years ago Sir Isaac Newton found the Spanish Mexican piece of eight to be of the intrinsic value of four shillings and six pence sterling.

The burden of a ship, as a weight, is ascertained by the depth of the water that she draws. On the principles of hydrostatics, the weight of any floating object is equal to that of the mass of water displaced by it: and the weight of a ship's burden is the difference between the column of water drawn by her when in ballast, and when laden. The draft of water, therefore, measured by the metre and its divisions, gives of itself the result, in tons of 1000 kilogrammes, by the mere multiplication of the dimensions of the vessels; the result giving cubic metres of water, each of which, saving the difference between the specific gravity of river or sea, and distilled water, will of course be of 1000 kilogrammes.

The size of casks was among the objects intended to be included in the reformed system: and regulations were adopted prescribing, first, that their dimensions should be of uniform proportions, the diameter of the two ends, that of the centre, and the length of the barrels, being as 8, 9, and 10½ to each other; and, secondly, that their contents should be in decimal or subdecimal divisions of *litres*. Tables were published prescribing the dimension in millimetres of the length and diameters of each cask, from the contents of 50 to those of 1000 litres. But the forms and proportions of casks are different in different countries, and in different places of the same country. These differences may arise from the nature of the substance, liquid or dry, which they are to contain; from the materials of which they are made or with which they are bound; from laws or usages long established, to which the cooper, the vintner, or brewer, the merchant, the miller, and other numerous professions dealing in articles which are packed in barrels, have accommodated themselves from time immemorial. With regard to articles of exportation, the laws of other countries also interpose, by prohibiting their admission in casks of other dimensions than those which have been used: and the instruction of 2 Frimaire, 11 (23d November, 1802), revoked the regulation of Pluviose, 7 (January, 1799), requiring only thenceforth, according to the proclamation of 11 Thermidor, 7 (29 July, 1799), that no wines or other liquors should be exposed to sale, unless branded with the mark of their contents in litres; with a recommendation, however, that casks should be made as much as possible in the dimensions and proportions which had been ordained in January, 1799.

The intentions of reformation upon the principles of uniformity and of decimal divisions were, in the novelty of the system, extended to the mariner's compass, which it was proposed to divide into forty rhumbs of wind, instead of thirty-two; to the log-line, the usual divisions of which are proportioned to the marine mile of sixty to a degree; to the sounding-line, which had usually been divided by French mariners, not into their fathoms of six, but into *brasses* of five royal feet; and to the cable's length, which was of 100 toises. Some of these were consequences of the project for dividing decimally time, and the quadrant of the circle: and the others followed from the substitution of the metre for the foot and toise.

The lapidaries and dealers in precious stones, throughout Europe, have a weight peculiar to themselves, under the denomination of the carat, which is nearly of the weight of three grains troy, and which they divide into halves, quarters, eighths, and sixteenths. As this trade is of extremely limited extent, even in Europe, it was to be considered only in the organization of a system for universal application.

It has been observed, that among the difficulties hitherto insuperable, which have opposed the establishment in fact of this system, thus apparently established by law, the most unmanageable of all has been found to be the adoption of the nomenclature. It is curious to observe the various expedients of legislation, to accommodate itself to the popular humors in this respect.

The law of the 1st of August, 1793, established all the principles of the new system, but under denominations different from those which had ever been used before, and not less different from those which have been adopted since. It directed the Academy of Sciences to compose an elementary book, containing a clear explanation of the new weights and measures, with tables of equalization, and instructions for adapting them to those which had been in use until then. A few days afterward the academy was itself abolished: but the duty of composing the book was assigned to a temporary commission, or board of weights and measures, consisting of the same persons who had been employed as members of the academy on the work. The book was composed and published in the year 1794. But, on the 19th of January of that year (30 Nivose, 2), the nomenclature had already changed; and, on the 7th of April, 1795 (18 Germinal, 3,) a nomenclature entirely new, with the exception of three or four words, was enacted. The names ordained by this law of 7th April, 1795, are still the proper technical appellations, and have already been mentioned, with their Greek and Latin prefixes of decimal multiples and subdivisions. The same law directed that weights or measures might be made of double, or of half the units and their tenth part, or tenth fold amounts; but that no other subdivision, or multiple, such as thirds, or quarters, or sixth, or eighth parts, should be allowed. The law of 19 Frimaire, 8 (10 December, 1799), declared the platina metre of 443,296 lines, and the kilogramme of 18,827.15 grains mark weight, to be definitive standard weight and measure; on the 13th Brumaire, 9 (4th November, 1800), the executive directory issued an arrêté, or order, authorizing, either in public writings or in habitual usage, what they called a translation into French words of the authentic nomenclature; so that the myriametre might be called a league, the kilometre a mile, the litre a pint, the kilogramme a pound, the hectogramme an ounce, the gramme a denier, and so of all the rest, excepting the *metre*, which was to have no synonymous or translated name, and the *stere*, for firewood and measures of solidity. This ordinance was never executed: and the minister of the interior, by an order of 30 Frimaire, 14 (21st December, 1855), directed all the subordinate administrations to use exclusively the denominations prescribed by the law of 7th April, 1795.

An imperial decree of 12th February, 1812, presents the subject under a new aspect, by ordaining,

1. That the *units* of weights and measures should remain unchanged, as established by the law of 10th December, 1799.

2. That the minister of the interior should cause to be made *instruments* for weight and mensuration, presenting the fractions or multiples of the said units the most commonly used in commerce, and accommodated to the wants of the people.

3. That these instruments should bear on their respective faces the comparison of the divisions and denominations established by law, with those which had been formerly used.

4. That after a term of ten years a report should be made to the emperor of the result of experience upon the improvements of which the system of weights and measures might be susceptible.

5. That in the mean time the legal system should continue to be taught in all the schools, and be exclusively used in all the public offices, and in all markets, halls, and commercial transactions.

For the execution and explanation of this decree, an ordinance was, on the 28th of March, 1812, issued by the minister of the interior, of the following purport:

Art. 1. Permission was granted to employ for the purpose of commerce,

1. A long measure equal to two metres, to be called a toise, and to be divided into six feet.

2. A measure equal to one-third of the metre, to be

called a foot, to be divided into twelve thumbs, and the thumb into twelve lines.

Each of these measures shall bear on one side the corresponding divisions of the metre, that is to say: the toise, two metres, divided into decimetres, and the first decimetre into millimetres; and the foot, three decimetres and one-third, divided into centimetres and millimetres, in all 333⅓ millimetres.

Art. 2. All cloths may be measured by a stick equal in length to twelve decimetres, to be called an *ell* (aune), which shall be divided into halves, quarters, eighths, and sixteenths, as well as into thirds, sixths, and twelfths. It shall bear on one of its sides the corresponding divisions of the metre, in centimetres only; that is to say, one hundred and twenty centimetres, numbered from ten to ten.

Art. 4. Corn and other dry measure articles may be measured, *in sales at retail*, by a vessel equal to one-eighth of the hectolitre, which shall be called a boisseau, and shall have its double, its half, and its quarter.

Art. 5. For *retail* sales of corn, seeds, meal, and roots, green or dry, the litre may be divided into halves, quarters, and eighths.

Art. 7. For *retail* sales of wine, brandy, and other liquors, measures of one-quarter, one-eighth, and one-sixteenth of the litre may be used; each of which measures shall be called by a name signifying its proportion to the litre.

Art. 8. For retail sales of all articles which are sold by weight, the shopman may employ the following *usual* weights:

The pound (livre), equal to half a kilogramme, or 500 grammes, which shall be divided into sixteen ounces.

The ounce (once), or sixteenth part of the pound, which shall be divided into eight gros.

The gros, or eighth part of the ounce, which shall be divided into halves, quarters, and eighths.

They shall bear, with their appropriate names, the indication of their weight in grammes, namely:

The pound - - - -	500 grammes
Half pound - - - -	250
Quarteron - - - -	125
Eighth, or ½ quarter - -	62.5
Ounce - - - - -	31.3
Half ounce - - - -	15.6
Quarter ounce, 2 gros - -	7.8
Gros - - - - -	3.9

And such is at this day the system of weights and measures, or, rather, such are the systems existing in France in their present condition; for, it cannot escape observation, that this decree and explanatory ordinance engraft upon the legal system an entirely new system, founded upon different, and in many important respects, opposite principles. So that the result hitherto of the most stupendous and systematic effort ever made by a nation to introduce uniformity in their weights and measures, has been a conflict between four distinct systems:

1. That which existed before the Revolution.

2. The temporary system established by the law of 1st August, 1793.

3. The definitive system established by the law of 10th December, 1799. And,

4. The *usual* system, *permitted* by the decree of 12th February, 1812.

This last decree is a compromise between philosophical theory and inveterate popular habits. Retaining the principle of decimal multiplication and division for the legal system, it abandons them entirely in the weights and measures which it allows the people to use. Instead of the metre and its decimals, it gives the people a toise of six feet, an aune of three feet and one-fifth, a foot of twelve thumbs, and a thumb of twelve lines. And these measures, instead of divisions exclusively decimal, are divisible in halves, thirds, quarters, sixths, eighths, twelfths, and sixteenths. Instead of a decimated kilogramme, it gives them a pound of sixteen ounces, an ounce of eight gros, and a gros of seventy-two grains. The measures of capacity, wet and dry, have the same indulgence: and while the standard weight and measure are deposited in the national archives, the people have restored to them for use all the names and divisions of their ancient weights and measures, though not the same things. For the toise, which is twice the length of the metre, is not the old toise; the foot, which is the third part of the metre, is not the *pied de roi:* but both are longer measures. The half kilogramme, which is a pound, is not the ancient mark-weight pound; nor are the boisseau or litre those of ancient times; they are all respectively near approximations to them.

If the existing system and practice terminated here, it would be far from having attained the ideal perfection of uniformity; but it is believed that, for a multitude of purposes, with this double and complicated system, there is yet a very extensive remnant in use of that which prevailed before the revolution. It appears, from questions at this time in discussion between the governments of the United States and of France, that the tonnage of the French shipping is calculated by admeasurements in cubic royal feet: and it appears hence probable that, in all the business of ship-building, and in practical navigation, those measures are still used. Without positive knowledge of the fact, the analogy of all experience warrants the conjecture, that in every part of France, remote from the capital, not only the use of the old legal system, but of the local weights and measures which prevailed in the various cities and districts of the country, is far from being eradicated.

The changes which have forced themselves upon the new system, under the attempt to reduce it to practice, should serve as admonitions to correct the errors of theory; but not operate as discouragement to the pursuit of the principal object, *uniformity*. The French metrology, in the ardent and exclusive search for an universal standard from nature, seems to have viewed the subject too much with reference to the nature of things, and not enough to the nature of man. Its authors do not appear to have considered, in all the bearings of the system, the proportions dictated by nature between the physical organization of man, and the *unit* of his weights and measures. The standard taken from the admeasurement of the earth had no reference to the admeasurement and powers of the human body. The metre is a rod of forty inches: and by applying to it exclusively the principle of decimal divisions, no measure corresponding to the ancient *foot* was provided. An unit of that denomination, though of slightly varied differences of length, was in universal use among all civilized nations: and the want of it is founded in the dimensions of the human body. Perhaps for half the occasions which arise in the life of every individual for the use of a linear measure, the instrument, to suit his purposes, must be portable, and fit to be carried in his pocket. Neither the metre, the half-metre, nor the decimetre, are suited to that purpose. The half-metre corresponds indeed with the ancient cubit: but perhaps one of the causes which have everywhere, since the time of the Greeks, substituted the foot in the place of the cubit, has been the superior convenience of the shorter measure. Besides which, the cubit being the unit, the half-cubit might serve the purposes of the foot; but the metre, divisible only by two and by ten, gave no measure practically corresponding with the foot whatever. It appears also not to have been considered, that decimal arithmetic, although affording great facilities for the computation of numbers, is not equally well suited for the divisions of material substances. A glance of the eye is sufficient to divide material substances into successive halves, fourths, eighths, and sixteenths. A slight attention will give thirds, sixths, and twelfths. But divisions of fifth and tenth parts are among the most difficult that can be performed without the aid of calculation. Among all its conveniences, the decimal division has the great disadvantage of being itself divisible only by the numbers two, and five. The duodecimal division, divisible by two, three, four, and six, would offer so many advantages over it, that while the French theory was in contemplation, the question was discussed, whether the reformation of weights and measures should not be extended to the system of arithmetic itself, and whether the number twelve should not be substituted for ten, as the term of the periodical return to the unit. Since the establishment of the French system, this idea has been reproduced by philosophical critics, as an objection against it; and Delambre, in the third volume of the Base du Système Metrique, p. 302, has considered it, and assigned the reasons for which it had been rejected. He admits, to the full extent, the advantages of a duodecimal over a decimal arithmetic; but alleges the difficulty of effecting the reformation, as the decisive reason against attempting it.

The review of the proceedings in Great Britain and France, relating to the uniformity of weights and measures, presents the general subject under two very different aspects, from the combination of which, it is believed, useful practical results may be derived. Considered as a whole, the established weights and measures of England are but the ruins of a system, the decays of which have been often repaired with materials adapted neither to the proportions, nor to the principles of the original construction. The metrology of France is a new and complicated machine, formed upon principles of mathematical precision, the adaptation of which to the uses for which it was devised is yet problematical, and abiding with questionable success the test of experiment.

The standard of nature of the English system is the length of the human foot, divided by the barley-corn. That of the French system is an aliquot part of the circumference of the earth decimally divided.

The material positive standard of the English system is an iron three-foot rod in the British exchequer. That of France is a platina metre in the national archives.

To the English system belong two different units of weight, and two corresponding measures of capacity, the natural standard of which is the difference between the specific gravities of wheat and wine. To the French system there is only one unit of weight and one measure of capacity, the natural standard of which is the specific gravity of water.

The French system has the advantage of unity in the weight and the measure, but has no common test of both. Its measure gives the weight only of water. The English system has the inconvenience of two weights and two measures; but each measure is at the same time a weight. Thus the gallon of wheat and the gallon of wine, though of different dimensions, balance each other as weights. A gallon of wheat and a gallon of wine, each, weigh eight pounds avoirdupois. This observation applies, however, only to the original principle of the English system, and not altogether to its present condition. This difference between the specific gravity of wheat and wine, is still the difference between the troy and avoirdupois weights, but not between the wine and corn gallons. A third vessel of capacity, for which neither the capacity nor the use is perceived, has usurped the place of the corn gallon; and it has been shown how it was introduced. The acts of parliament prescribing the dimensions of the bushel and of the wine gallon in cubic inches, have assumed them from existing standards, or erroneous calculations: and the proportions between the measures of corn and of wine,

which belonged originally to the system, are now transferred to those of the wine and beer, for which, if the reason was that beer being a home-made liquor and wine a
foreign production, beer a comfort of the poor, and wine a
luxury of the rich, the former ought to be dealt out in
larger portions, and more lightly touched with taxation,
it proceeded from the best motives of political morality;
but which might have been as well accomplished by reducing the tax as by enlarging the measure. As vessels of
capacity for fluids, there can be no useful reason for different measure, except the proportion of specific gravities.

In the English system, the smaller of the two weights
was originally also identical with the coin: a pound of the
weight was a pound sterling in silver money. But this
property it has irrecoverably lost.

In the French system, the weight is not a coin; but the
metallic coins are weights. Gold, silver, mixed metal, and
copper, are all coined in proportions of weight and relative
value prescribed by law.

In our monetary system, we have discarded the last
trace of identity between weights and coins, by ceasing to
apply to money the name of pound or penny. Our coins
are of prescribed weight and purity, but in no convenient
or uniform proportions to each other.

In the English system the two weights are standards of
verification to each other; the two pounds being in the
proportion to each other of 144 to 175, and the pound
avoirdupois being of 7,000 grains troy. For quantities
amounting to one-fourth of a hundred pounds or more,
the English avoirdupois weight requires an accession of
12 per cent.; 28 pounds pass for 25, 56 for 50, and 112 for

100. The original motive for this must have been the convenience of dividing the hundred into halves, quarters, eighths, and sixteenths, without making fractions of a pound. The true hundred can thus be divided into no whole number less than a quarter, or 25.

In the English system, the standard linear measure is connected with the weights by the specific gravity of spring-water, of which a measure of one cubic foot contains one thousand ounces avoirdupois.

In the French system, the standard linear measure is connected with the weight and the measure of capacity, by the specific gravity of distilled water, at its greatest density, one cubic decimetre of such water being the weight of the kilogramme, and filling the measure of the litre.

In the English system, every weight and every measure is divided by different and, seemingly, arbitrary numbers; the foot into twelve inches; the inch, by law, into three barley-corns, in practice sometimes into halves, quarters, and eighths, sometimes into decimal parts, and sometimes into twelve lines; the pound avoirdupois into sixteen ounces, and the pound troy into twelve, so that while the pound avoirdupois is heavier, its ounce is lighter than those of the troy weight. The ton, in the English system, is both a weight and a measure. As a measure, it is divided into four quarters, the quarter into eight bushels, the bushel into four pecks, etc. As a weight, it is divided into twenty hundreds, of 112 pounds, or 2,240 pounds avoirdupois. The gallon is divided into four quarts, the quart into two pints, and the pint into four gills.

In the French system, decimal divisions were prescribed

by law exclusively. The binary division was allowed, as being compatible with it: but all others were rigorously excluded; no thirds, no fourths, no sixths, no eighths, or twelfths. But this part of the system has been abandoned: and the people are now allowed all the ancient varieties of multiplication and division, which are still further complicated by the decimal proportions of the law.

The nomenclature of the English system is full of confusion and absurdity, chiefly arising from the use of the same names to signify different things; the term pound to signify two different weights, a money of account, and a coin; the gallon and quart to signify three different measures; and other improper denominations constantly opening avenues to fraud.

The French nomenclature possesses uniformity in perfection, every word expressing the unit weight or measure which it represents, or the particular multiple or division of it. No two words express the same thing: no two things are signified by the same word.

If, with a view to fixing the standard of weights and measures for the United States, upon the principles of the most extensive uniformity, the question before Congress should be upon the alternative, either to adhere to the system which we possess, or to adopt that of France in its stead, the first position which occurs as unquestionable is, that change, being itself diversity, and therefore the opposite of uniformity, cannot be a means of obtaining it, unless some great and transcendent superiority should demonstrably belong to the new system to be adopted, over the old one to be relinquished.

In what then does the superiority of the French system,

in all its novelty and freshness, over that of England, in
all its decays, theoretically consist?

1. In an invariable standard of linear measure, taken
from nature, and being an aliquot decimal portion of the
quarter of the meridian.

2. In having a single unit of all weights, and a single
unit of measures of capacity for all substances, liquid or
dry.

3. In the universal application of the decimal arithmetic, to the multiples and divisions of all weights and
measures.

4. In the convenient proportions by which the coins
and money of account are adjusted to each other and to
the weights.

5. In the uniformity, precision, and significancy of the
nomenclature.

1. If the project of reforming weights and measures had
extended, as was proposed by the French system, to the
operations of astronomy, geography, and navigation; if
the quadrant of the circle and of the sphere had been divided into one hundred degrees, each of one hundred
thousand metres; the assumption of that measure would
have been an advantage much more important than it is,
or can be, in the present condition of the system. Whether
it would have compensated for disturbing that uniformity
which exists, and which has invariably existed, of the division into ninety degrees, with sexagesimal subdivisions of
minutes and seconds, is merely matter of speculation. At
least, it has been found impracticable, even in France, to
carry it into effect: and without it, the metre, as the
natural standard of the system, has no sensible advantage

over the foot. To a perfect system of uniformity for all weights and measures, as an aliquot part of the circumference of the earth is not only a better natural standard unit than the pendulum, or the foot, but it is the only one that could be assumed. Every voyage round the earth is an actual mensuration of its circumference. All navigation is admeasurement: and no perfect theory of weights and measures could be devised, combining in it the principle of decimal computation, of which any other natural standard whatever could accomplish the purpose. Its advantages over the pendulum are palpable. The pendulum bears no proportion to the circumference of the earth, and cannot serve as a standard unit for measuring it. Yet a system of weights and measures, which excludes all geography, astronomy, and navigation, from its consideration, must be essentially defective in the principle of uniformity.

But, if the metre and its decimal divisions are not to be applied to those operations of man, for which it is most especially adapted; if those who circumnavigate the globe in fact are to make no use of it, and to have no concern in its proportions; if their measures are still to be the nonagesimal degree, the marine league, the toise, and the foot; it is surely of little consequence to the farmer who needs a measure for his corn, to the mechanic who builds a house, or to the townsman who buys a pound of meat, or a bottle of wine, to know that the weight, or the measure which he employs, was standarded by the circumference of the globe. For all the uses of weights and measures, in their ordinary application to agriculture, traffic, and the mechanic arts, it is perfectly immaterial what the natural standard, to which they are referable, was. The foot of

Hercules, the arm of Henry the First, or the barley-corn, are as sufficient for the purpose as the pendulum, or the quadrant of the meridian. The important question to them is, the correspondence of their weight or measure with the positive standard. With the standard of nature, from which it is taken, they have no concern, unless they can recur to it as a test of verification. However imperfect for this end the human foot, or the kernel of wheat or barley, may be, they are at least easily accessible. It is a great and important defect of the systems which assume the meridian or the pendulum for their natural standard, that they never can be recurred to without scientific operations.

This is one great advantage which a natural standard, taken from the dimensions and proportions of the human body, has over all others. We are perhaps not aware how often every individual, whose concerns in life require the constant use of long measures, makes his own person his natural standard, nor how habitually he recurs to it. But the habits of every individual inure him to the comparison of the definite portion of his person, with the existing standard measures to which he is accustomed. There are few English men or women but could give a yard, foot, or inch measure, from their own arms, hands, or fingers, with great accuracy. But they could not give the metre or decimetre, although they should know their dimensions as well as those of the yard and foot. When the Russian General Suwarrow, in his Discourses under the Trigger, said to his troops, "a soldier's step is an arsheen;" he gave every man in the Russian army the natural standard of the long measure of his country. No Russian soldier could ever afterward be at a loss for an arsheen. But, although it is precisely twenty-eight English inches, being otherwise divided, a Russian soldier would not, without calculation, be able to tell the length of an English yard or inch.

Should the metre be substituted as the standard of our weights and measures, instead of the foot and inch, the natural standard which every man carries with him in his own person would be taken away ; and the inconvenience of the want of it would be so sensibly felt, that it would be as soon as possible adapted to the new measures: every man would find the proportions in his own body corresponding to the metre, decimetre, and centimetre, and habituate himself to them as well as he could. If this conjecture be correct, is it not a reason for adhering to that system which was founded upon those proportions, rather than resort to another, which, after all, will bring us back to the standard of nature in ourselves.

2. The advantage of having a single unit of all weights and a single unit of measures of capacity, is so fascinating to a superficial view, that it would almost seem presumption to raise a question, whether it be so great as at first sight it appears. The relative value of all the articles which are bought and sold by measures of capacity, is a complicated estimate of their specific gravity and of the space which they occupy. If both these properties are ascertained by one instrument for any one article, it cannot be applied with the same effect to another. Thus the litre, in the French system, is a measure for all grains and all liquids : but its capacity gives a weight only for distilled water. As a measure of corn, of wine, or of oil, it gives the space which they occupy, but not their weight. Now,

as the weight of those articles is quite as important in the
estimate of their quantities, as the space which they fill,
a system which has two standard units for measures of
capacity, but of which each measure gives the same weight
of the respective articles, is quite as uniform as that which,
of any given article, requires two instruments to show its
quantity; one to measure the space it fills, and another
for its weight. It has been observed, that nature, in the
relations which she has established between man and the
earth upon which he dwells, and in providing for the
wants resulting to him from these relations, offers him in his
own person two natural standards even of linear measure;
one for the range of his own movements upon the earth,
and the other for articles loosened from the earth, and
which are adapted to the immediate wants of his person.
He finds by experience that these may with increased convenience be reduced to one. It is not exactly so with
weights or measures of capacity. From the moment when
man becomes a tiller of the ground, and civil society is
organized; from the moment when the mutual exchange
between the wants of one and the superfluities of another
commences; measures of capacity and weights are necessary to the operation. The use of metals as common
standards of value, is of later origin, and when first
applied to that purpose, they are always delivered by
weight. The first and most important article of traffic is
corn, the first necessary of life: wine and oil successively
come next: milk and honey follow. For all these, weights
and measures of capacity are indispensable. When the
metals are first used as common instruments of exchange,
the proportions of their qualities are estimated by their

weight. But that weight could not be ascertained by itself. The metal being in one scale, there must be something else to balance it in the other: and that other substance, first of all, would, whenever it should have come into use for food, be corn. It might next be wine. But thus compared, it would immediately be seen that the vessel containing of wine a counterpoise to the given 'metallic weight, would not contain a counterpoise of wheat to the same weight: and what could more naturally suggest itself than the device, to bring to the scales the wheat in a measure to balance the weight, and the wine in a measure to produce the same effect? The metallic weight would then become the common standard for both, but would neither be the same weight by which its own gravity had been ascertained, nor a substitute for it. Thus, the operation of weighing implies in its nature the use of two articles, each of which is the standard testing the gravity of the other. And in the difference between the specific gravities of corn and wine, nature has also dictated two standard measures of capacity, each of them equiponderant to the same weight.

This diversity existing in nature, the troy and avoirdupois weights, and the corn and wine measures of the English system, are founded upon it. In England it has existed as long as any recorded existence of man upon the island. But the system did not originate there, neither was Charlemagne the author of it. The weights and measures of Rome and of Greece were founded upon it. The Romans had the *mina* and the *libra*, the nummulary pound of twelve ounces, and the commercial pound of sixteen. And the Greeks, as well as the Romans, had a weight for small and precious, and a weight for bulky and cheap commodities. The Greeks denominated them by significant terms, *the weight for measure* and *the weight for money*. Whether the ounce, of which these pounds were composed, was the same, is a subject of much controversy, but of little importance to decide. At the period of the lower empire, these two weights were known by the name of the *eastern* and *western* pound. And the denomination of the former was the same in England: it was the *easterling* pound, and the origin of the term sterling in the English language: it was the pound of the eastern nations, by which Europe was overrun in the decline of the Roman Empire. The avoirdupois pound had the same origin: for it came through the Romans from the Greeks, and through them, in all probability, from Egypt. Of this there is internal evidence in the weights themselves, and in the remarkable coincidence between the cubic foot and the thousand ounces avoirdupois, and between the ounce avoirdupois and the Jewish silver shekel. The Greek foot was within a fraction of less than the hundredth part of an inch, the same with that of England. The ounce avoirdupois is the same with the Roman and Attic ounce, and the exact double of the Jewish shekel. The Silian plebiscitum, or ordinance of the Roman people, of the year 509, two hundred and fifty years before the Christian æra, declares, that a quadrantal of wine shall be eighty pounds, a congius of wine ten pounds; that six sextarii make a congius of wine; forty-eight sextarii a quadrantal of wine; that the sextarius of liquid and dry measures should be the same; and that sixteen pounds make the modius. The congius was the Roman gallon, and the modius the Roman peck. The quadrantal was the same as the amphora, and was formed from the cubic foot of water, so that eighty pounds of wine were equal to a cubic foot of water.

The same combinations are traced with equal certainty to the Greeks and Egyptians: and, if the shekel of Abraham was the same as that of his decendants, the avoirdupois ounce may, like the cubit, have originated before the flood.

This diversity is, therefore, founded in the nature of things; and may be stated by the following rule: that whatever is sold *by* weight, in *measure* must have a measure for itself, which will serve for no other article, of different specific gravity; and as wheat and wine are both articles of that description, as their specific gravities are very materially different, although they are very suitable to be weighed by the same weight, they yet require different measures, to place them in equipoise with that weight. The difference of specific gravity between the vinous and watery fluids is so slight, that neither in the Greek, the Roman, nor the English system, was there any account taken of it. But with regard to oil, it appears that the Greeks had a separate measure adapted to its specific gravity, which they considered as being in proportion to that of wine or water as nine to ten.

Notwithstanding, therefore, the first appearance of superior uniformity and simplicity presented by the single unit of weights, and single measure of capacity in the new system of France, it appears to be more conformable to the order of nature, and more subservient to the purposes of man, that there should be two scales of weight and two measures of capacity, graduated upon the respective specific gravities of wheat and wine, than with a single weight and a single measure, to be destitute of any indication of weight in the measure.

This conclusion has been confirmed by a very striking fact, which has occurred in France under the new system. By an ordinance of police, approved on the 6th of December, 1808, by the Minister of the Interior, it is prescribed that the sale of oil in Paris by retail shall be *by* weight, in measures, containing five hectogrammes, one double hectogramme, one hectogramme, etc. And these measures being cylinders of tin, are stamped with initial letters, indicating that one is for sweet-oil, and the other for lamp oil. So that here are two new measures of capacity altogether incongruous to the new system, each differing in cubic dimensions from the other, though to measure the single article of oil, and both differing from the litre. They attach themselves indeed to the new system by *weight*, but abandon entirely its pretensions to unity of measure; and fall at once into the principle of the old system, of adapting the measure to the weight.

By the usages of modern times, the weight of wine is of little or no consideration. Its first admeasurement is in casks, of different dimensions in different places, and which cannot be made uniform, unless by a system of metrology common to many nations. It is sold wholesale by the cask or hogshead, the contents of which are ascertained by mechanical gauging instruments, adapted to the smaller measures of capacity of the country where it is to be consumed. These instruments give the solid contents of the vessel, and the number of the standard measures of the country which it contains. The gauging-rods used in

England and the United States give the contents in cubic inches and wine gallons. As a test of the quantity of wine contained in the cask, this mode of admeasurement is less certain and effectual than weight, especially if the cask is not full: but, being more convenient and easy of application, and specially adapted to the legal measure of the gallon in cubic inches, it has superseded altogether the use of weights as proofs of the quantity of wine. By retail, the article is sold either in the gallon measures fixed by law at 231 cubic inches, or in bottles of no definite measure, but containing an approximation to a quart or pint.

Our system of weights and measures, by the substitution of the wine gallon of 231 for that of 224 cubic inches, has lost the advantage which it originally possessed of testing the accuracy of a wine measure by its weight. The average specific gravity of wine is of 250 grains troy weight to a cubic inch: four inches therefore make a thousand grains, and twenty-eight inches a pint weighing one pound avoirdupois. These coincidences would be of great utility and convenience, and would be rendered still more so by another, which is, that this number of 224 inches is the exact decimal part of 2240, the number of pounds avoirdupois that go to a ton. As it now exists, therefore, the measure of the gallon of wine does not show its weight; and the unity of the measure of capacity in the French system, is an advantage not compensated by any benefit derived from the different dimensions of our corn and wine gallons.

Our country is not as yet a land of vineyards. We have no "flowery dales of Sibma, clad with vines." Wine is an article of importation; an article of luxury, in a great measure confined to the consumption of the rich. Its distribution in measure, and the exactness of the measure by which it is distributed, is not an incident which every day comes home to the interests and necessities of every individual. We have less reason for regretting, therefore, the loss of a measure which would prove its integrity by the weight; and more reason for preferring the uniformity of singleness in the French system of capacious measures, to the uniformity of proportion which belonged originally to the English. That proportion itself we have lost by the establishment of a wine gallon of 231, and a corn bushel of 2,150 cubic inches: and although it exists in the troy and avoirdupois weights, and in the wine and beer gallons, it exists to none of the useful purposes for which it was originally intended, and to which in former days it was turned.

The consumption of wine in modern times is exceedingly diminished, not only by the substitution of beer, and of spirits distilled from grain in the countries where the wine is not cultivated, but by the use, now become universal, of decoctions from aromatic herbs and berries. Tea and coffee are potations unknown to the European world until within these two centuries: and they have probably diminished by one-half the consumption of wine throughout the world.

The measures, by which solid and liquid substances are sold, are not and cannot conveniently be the same. The form and the substances of the vessels in which they are kept are altogether different. Grain is usually kept in bags, until ground into meal. Liquids, in large quantities, are kept in wooden vessels of peculiar construction, founded upon the properties of fluids and the laws of hydrostatics: in small quantities, they are kept in vessels of glass, adapted by their form to the facility of pouring them off without loss. Such vessels are utterly unsuitable for containing grain, or any other solid substance. The forms, both of casks and of bottles, are among the most difficult forms into which cubical extension can be moulded for ascertaining quantity by linear measure. They not only contained the problem, hitherto unsolvable to man, of squaring the circle; but some of the most recondite mysteries of the conic sections. They are neither cylinders, nor ellipses, nor cones, nor spheres; but a combination of all these forms. Grain may be measured by a cylindrical or a cubical vessel, at pleasure. The cylindrical form is best adapted to convenience; and by the known proportion of the diameter to the circle, its solid contents in linear measure may be ascertained with sufficient accuracy and little difficulty. Grain cannot be kept in vessels with large bodies, long necks, and narrow mouths. Liquids can be well kept for preservation in no other. Grain is a swiftly perishable substance, which must generally be consumed within a year from its growth: wines and spirituous liquors in general may be kept many years, and the vessels in which they are kept must be of forms and substances calculated to guard against loss by evaporation, fermentation, or transudation. So different indeed are all the properties of grain and of all liquids, that, instead of requiring the same measure to indicate their qualities, the call of nature is for different vessels, of different substances, and in different forms.

The most certain and convenient test for the accuracy of dry measures is linear measure; that of liquids is *weight*. The sextarius of the Roman system, and the litre of the French, were measures common both to wet and dry substances. But in applying it, the Romans formed a liquid measure of ten pounds *weight*, and a dry measure of sixteen. The French litre combines both the tests of linear measure and of weight for the single article of distilled water, at a certain temperature of the atmosphere: but it is not the test of weight for anything else. The hectolitre of wine or of corn is no indication of the weight of either. The sale of wheat, from the nature of the article, must usually be in large quantity, seldom less than a bushel. The unit of the measure declared in Magna Charta is the quarter of a ton, or eight bushels. Wine is an article the sale of which is as frequent in retail as by wholesale. The accuracy of its admeasurement in small quantities is important. In this respect it has an analogy to the precious metals. In fine, the purchase and sale of liquid and dry substances is, by the constitution of human society, not at the same times, or places, nor by the same persons. Their difference in the origin is that of the vineyard and the cornfield. They pass thence respectively to the wine-press and the flour-mill; thence to the vintner and the flour merchant, in vessels already adapted to their respective conditions; the corn having undergone a transformation requiring a different measure from the wheat. Trace them through all their meanderings in the circulation of civil society, till they come to their common ultimate use for the subsistence of man; it will never be found that the same measures are necessary for, or suitable to, them. · The wheat comes in the shape of meal or bread, to be measured by weight; and the liquors in casks or bottles, and still in the form given to them by distillation. The distiller and the brewer, who manufacture the liquid from the grain, have occasion for both measures; but the articles come to them in one form, and go from them in the other; nor is there any apparent necessity that they should receive and issue them by the same measure.

There are conveniences in the intercourse of society, connected with the use of smaller and more minutely perfect weights and measures of capacity, for sales of articles by retail, than by wholesale, and for articles of great price though of small bulk. Thus, drugs, as articles of commerce, and in gross, are sold by the avoirdupois or commercial pound; used as medicines, in minute quantities, and compounded by the apothecary, they are sold by the smaller or nummulary weight. The laws of Pennsylvania authorize innkepers to sell beer, *within the house*, by the wine measure; but, for that which they send out of the house, require them to use the beer gallon or quart. In both these cases the difference of the measure forms part of the compensation for the labor and skill of the apothecary, and part of the profits necessary to support the establishment of the publican. There is, finally, an important advantage in the establishment of two units of weights and of measures of capacity, by the possession in each of a standard for the verification of the other. It serves as a guard against the loss or destruction of the positive standard of either. The troy and avoirdupois pounds are to each other as 5,760 to 7,000. Should either of these standard pounds be lost, the other would supply the means of restoring it. The same thing might be effected by the measures of beer and of wine. The French system has designated the pendulum as such a standard for the verification of the metre. The English system gives, in each weight and measure, a standard for the other.

The result of these reflections is, that the uniformity of nature for ascertaining the quantities of all substances, both by gravity and by occupied space, is a uniformity of proportion, and not of identity; that, instead of one weight and one measure, it requires two units of each, *proportioned* to each other; and that the original English system of metrology, possessing two such weights, and two such measures, is better adapted to the only uniformity applicable to the subject, recognized by nature, than the new French syetem, which, possessing only one weight and one measure of capacity, identifies weight and measure only for the single article of distilled water; the English uniformity being relative to the *things* weighed and measured, and the French only to the instruments used for weight and mensuration.

3. The advantages of the English system might, however, be with ease adapted to that of France, but for the exclusive application in the latter of the decimal arithmetic to all its multiples and subdivisions. The decimal numbers, applied to the French weights and measures, form one of its highest theoretic excellences. It has, however, been proved by the most decisive experience in France, that they are not adequate to the wants of man in society: and, for all the purposes of retail trade, they have been formally

abandoned. The convenience of decimal arithmetic is in
its nature merely a convenience of calculation: it belongs
essentially to the keeping of accounts; but is merely an
incident to the transactions of trade. It is applied, therefore, with unquestionable advantage, to moneys of account,
as we have done: yet, even in our application of it to the
coins, we have not only found it inadequate, but in some
respects inconvenient. The divisions of the Spanish
dollar, as a coin, are not only into tenths, but into halves,
quarters, fifths, eighths, sixteenths, and twentieths. We
have the halves, quarters, and twentieths, and might have
the fifths; but the eighth makes the fraction of the cent,
and the sixteenth even a fraction of a mill. These eighths
and sixteenths form a very considerable proportion of our
metallic currency: and although the eighth dividing the
cent only into halves adapts itself without inconvenience
to the system, the fraction of the sixteenth is not so tract-
able; and in its circulation, as small change, it passes for
six cents, though its value is six and a quarter, and there is ·
a loss by its circulation of four per cent, between the buyer
and the seller. For all the transactions of retail trade, the
eighth and sixteenth of a dollar are among the most useful
and convenient of our coins: and, although we have never
coined them ourselves, we should have felt the want of
them, if they had not been supplied to us from the coinage
of Spain.

This illustration, from our own experience, of the modification with which decimal arithmetic is adaptable even
to money, its most intimate and congenial natural relative,
will disclose to our view the causes which limit the exclusive application of decimal arithmetic to *numbers,* and

admit only a partial and qualified application of them to weight or measure.

It has already been remarked, that the only apparent advantage of substituting an aliquot part of the circumference of the earth, instead of a definite portion of the human body, for the natural unit of linear measure, is, that it forms a basis for a system embracing *all* the objects of human mensuration; and that its usefulness depends upon its application to geography and astronomy, and particularly to the division of the quadrant of the meridian into centesimal degrees. In the novelty of the system, this was attempted in France, as well as the decimal divisions of time, and of the rhumbs of the wind. A French navigator, suffering practically under the attempt thus to navigate, decimally, the ocean, recommended to the national assembly to decree, that the earth should perform four hundred revolutions in a year. The application of decimal divisions to time, the circle, and the sphere, are abandoned even in France. And for all the ordinary purposes of mensuration, excepting itinerary measure, the metre is too long for a standard unit of nature. It was a unit most especially inconvenient as a substitute for the foot, a measure to which, with trifling variations of length, all the European nations and their descendants were accustomed. The foot rule has a property very important to all the mechanical professions, which have constant occasion for its use: it is light, and easily portable about the person. The metre, very suitable for a staff, or for measuring any portion of the earth, has not the property of being portable about the person: and, for all the professions concerned in ship or house building, and for all who have occasion to use mathematical instruments, it is quite unsuitable. It serves perfectly well as a substitute for the yard or ell, the fathom or perch ; but not for the *foot*. This inconvenience, great in itself, is made irreparable when combined with the exclusive principle of decimal divisions. The union of the metre, and of decimal arithmetic, rejected all compromise with the foot. There was no legitimate extension of matter intermediate between the ell and the palm, between forty inches and four. This decimal despotism was found too arbitrary for endurance: not only the foot, but its duodecimal divisions, were found to be no arbitrary or capricious institutions, but founded in the nature of the relations between man and things. The duodecimal division gives equal aliquot parts of the unit, of two, three, four, and six. By giving the third and the fourth, it indirectly gives the eighth and sixteenth, and gives facility for ascertaining the ninth, or third of the third. Decimal division, in giving the half, does not even give the quarter, but by multiplication of the subdivisions. It is incommensurable with the *third*, which unfortunately happened to be the foot, the universal standard unit of the old metrology. The choice of the kilogramme, or cubical decimetre of distilled water, as the single standard unit of weights, with the application to it of the decimal divisions, was followed by similar inconveniences. The pound weight should be a specific gravity easily portable about the person, not only for the convenience of using it as an instrument, but as the measure of quantities to be carried. To the common mass of the people, the use of weights is in the market or the shop. The article weighed is to be carried home. It is an article of food for the daily subsistence of

the individual or his family. As he has not the means of
purchasing it in large quantities, it must often be sold in
quantities represented by the pound weight, which, like the
foot rule, with various modifications, is universally used
throughout the European world. Subdivisions of that
weight, the half, the quarter of a pound, are often necessary to conciliate the wants and the means of the neediest
portion of the people; that portion to whom the *justice* of
weight and measure is a necessary of life, and to whom it
is one of the most sacred duties of the legislator to secure
that justice, so far as it can be secured by the operation of
human institutions. The half of the kilogramme was
nearly equivalent to the ancient Paris pound. But there
was in the new system no half or quarter of a pound, because there was no quarter or eighth of a kilogramme.
There was no intermediate weight between the pound or
half-kilogramme and the hectogramme, which was a fifth
part of a pound.

The *litre,* or unit of measures of capacity in the new
system, had one great advantage over the linear and weight
units, by its near equivalence to the old Paris pint, of which
it was to take the place. But on the other hand, decimal
divisions are still more inapplicable to measures of capacity
for liquids than to linear measures or weights. The substance in nature best suited for a retail measure of liquids
is tin: and the best form in which this measure can be
moulded is a slight approach from the cylinder to the cone.
Our quart and gallon wine and beer measures are accordingly of that form, as are all the most ordinary vessels used
for drinking. In the new French system, the form of all
the measures of capacity is cylindrical; and the litre is a

measure, the diameter of which is half its depth. It is, therefore, easily divisible into halves, quarters, and eighths; for it needs only thus to divide the depth, retaining the same diameter. But all conveniences of proportion are lost by taking one-tenth of the depth and retaining the same diameter: and if the diameter be reduced, there is no means other than complicated calculation, squarings of the circle, and extractions of cube roots, that will give one liquid measure which shall be the tenth part of another.

In the promiscuous use of the old weights and measures and the new, which was unavoidable in the transition from the one to the other, the approximation to each other of the quarter and the fifth parts of the unit became a frequent source of the most pernicious frauds; frauds upon the scanty pittance of the poor. The small dealers in groceries and liquors, and marketmen, gave the people the fifth of a kilogramme for a half-pound, and a fifth of the litre for a half-setier. The most easy and natural divisions of liquids are in continual halvings; and the Paris pint was thus divided into halves, quarters, eighths, sixteenths, and thirty-second parts, by the name of chopines, half-setiers, possoms, half-possoms, and roquilles. The half-setier, just equivalent to our half-pint, was the measure in most common use for supplying the daily necessities of the poor; and thus the decimal divisions of the law became snares to the honesty of the seller, and cheats upon the wants of the buyer.

Thus, then, it has been proved, by the test of experience, that the principle of decimal divisions can be applied only with many qualifications to any general system of metrology; that its natural application is only to numbers; and that

time, space, gravity, **and** extension inflexibly reject its
sway. The new metrology of France, after trying it in its
most universal theoretical application, has been compelled
to renounce it for all the measures of astronomy, geography,
navigation, time, the circle, and the sphere; to modify it
even for superficial and cubical linear measure, and to compound with vulgar fractions in the most ordinary and daily
uses of all its weights and all its measures. It has restored
the foot, the pound, and **the** pint, with all their old subdivisions, though not exactly with their old dimensions.
The foot, with its duodecimal divisions into thumbs and
lines, returns in the form the most irreconcileable possible,
with the decimals of the metre; for it comes in the proportion of three to ten, and consists of $333\frac{1}{3}$ millimetres. This
indulgence to linear measure is without qualification, and
may be used in all commerce, whether of wholesale or
retail. The restoration of the pound, the boisseau,* and
the pint, is limited to retail trade. The fractions of the
pound are as averse to decimal combinations as those of the
foot. The eighth of a pound, for instance, is 625 decigrammes, each of about $1\frac{1}{2}$ grains troy weight. The half
of this eighth is an *ounce*, to form which, decimally, re-

---

* One of the most abundant sources of error and confusion in relation to weights and measures, arises from mistranslation of those of
one country into the language of another. Thus, to call the *pinte* of
Paris a pint, is to give an incorrect idea of its contents. The Paris
pinte corresponded with our wine quart, containing 46.95 French,
or 58.08 English cubic inches. To call the boisseau a *bushel* is a
still greater incongruity between the word and the idea connected
with it. The *boisseau* contained 655 French cubic inches, and was
less than $1\frac{1}{2}$ pecks English. The minot, or three boisseaus, was the
measure corresponding with the English bushel.

quires a recourse to another fractional stage, and to say 312.5 milligrammes. But the milligramme, being equivalent to less than $\frac{1}{4}$ of a grain troy weight, is too minute for accurate application, so that it is called, and marked upon the weight itself, as 31.3 decigrammes. The half-ounce, instead of 1562.5 decimilligrammes, is marked for 15.6 decigrammes. The quarter of an ounce, instead of 7.8125, passes for 7.8 decigrammes, and the gros, or groat, instead of 3.90625, is abridged to 3.9. The ounce and all the smaller weights, therefore, reject the coalition of subdivision by decimal and vulgar fractions, and the weights for account are different from the weights for trade.

From the verdict of experience, therefore, it is doubtful whether the advantage to be obtained by any attempt to apply decimal arithmetic to weights and measures, would ever compensate for the increase of diversity which is the unavoidable consequence of change. Decimal arithmetic is a contrivance of man for computing numbers; and not a property of time, space, or matter. Nature has no partialities for the number ten: and the attempt to shackle her freedom with them, will forever prove abortive.

The imperial decree of March, 1812, by the reservation of a purpose to revise the whole system of the new metrology, after a further interval of ten years of experience, seems to indicate a doubt, whether the system itself can be maintained. Ten years from 1812, was a period far beyond that which Providence had allotted to the continuance of the imperial government itself. The royal government of France, which has since succeeded, has hitherto made no change in the system. Whether, at the expiration

of the ten years, limited in the decree, the proposed revisal
of it will be accomplished by the present government, is
not ascertained. In the mean time, the whole system
must be considered as an experiment upon trial even in
France: and should it ultimately prove, by its fruits,
worthy of the adoption of other nations, it will at least be
expedient to postpone engrafting the scion, until the character of the tree shall have been tested, in its native soil,
by its fruits.

4. The fourth advantage of the French metrology over
that which we possess, consists in the convenient proportions, by which the coins and moneys of account are
adjusted to each other, and to the weights.

This is believed to be a great and solid advantage; not
possessed exclusively by the French system, for it was, in
high perfection, a part of the original English system of
weights and measures, as has already been shown. It was
more perfect in that system, because the silver coins and
weights were not merely proportioned to each other, but
the *same.* This is not the case with the French coins;
and even their *proportions* to the weights are disturbed
and unhinged by the mint allowance, or what they call
toleration of inaccuracy, both of weight and alloy. This
toleration, which is also technically called the *remedy,*
ought everywhere to be exploded. It is in no case necessary. The toleration is injustice: the remedy is disease.
If it were the duty of this report to present a system of
weights, measures, and coins, all referable to a single
standard, combining with it, as far as possible, the decimal arithmetic, and of which uniformity should be the
pervading principle, without regard to existing usages, it

would propose a silver coin of nine parts pure and one of
alloy: of thickness equal to one-tenth part of its diameter;
the diameter to be one-tenth part of a foot, and the foot
one-fourth part of the French metre. This dollar should
be the unit of weights as well as of coins and of accounts;
and all its divisions and multiples should be decimal. The
unit of measures of capacity should be a vessel containing
the weight of ten dollars of distilled water, at the temperature of ten degrees of the centigrade thermometer: and
the cubical dimensions of this vessel should be ascertained
by the weight of its contents; the decimal arithmetic
should apply to its weight, and convenient vulgar fractions to its cubical measure. This system once established,
the standard weight and purity of the coin should be
made an article of the constitution, and declared unalterable by the legislature. The advantage of such a system
would be to embrace and establish a principle of uniformity with reference to time, which the French metrology
does not possess. The weight would be a perpetual guard
upon the purity and value of the coin. No second weight
would be necessary or desirable. The coin and the weight
would be mutual standards for each other; accessible, at
all times, to every individual. Should the effect of such a
system only be, as its tendency certainly would be, to de-
prive the legislative authority of the power to debase the
coins, it would cut up by the roots one of the most pernicious practices that ever afflicted man in civil society.
By its connection of the linear standard with the French
metre, it would possess all the advantages of having that
for a unit of its measures of length, and a link of the most
useful uniformity with the whole French metrology.

But the consideration of the coins is beyond the scope of the resolutions of the two Houses; nor is their relation to the weights and measures of the country viewed, by the constitution and laws of the United States, as that of parts of one entire system. Excepting the application of decimal divisions to our money of account, and the establishment of the dollar as the unit both of the money of account and of the silver coins, our moneys have no uniform or convenient adjustment to our weights. The proportion of alloy is not the same in our coins of silver, as in those of gold: and the only connection between our monetary system and our weights and measures is, that the gravity and proportional purity of the coins is prescribed in troy weight grains. To obtain, therefore, the advantage existing in the French metrology, of easy proportions between the weights and coins, or the still greater advantage of identity between them which belonged to the old English system, an entire change would be necessary in the fabrication of our coins, and in our moneys of account. It is, at least, extremely doubtful whether the benefits to be derived from such a change would be equivalent to the difficulties of achieving it, and the hazard of failing in the attempt.

5. The last superior advantage of the French metrology is, the *uniformity*, precision, and significancy, of its nomenclature.

In mere speculative theory, so great and unequivocal is this advantage, that it would furnish one of the most powerful arguments for adopting the whole system to which it belongs. In every system of weights and measures, ancient or modern, with which we are acquainted, until the new system of France, the poverty and imperfection of language has entangled the subject in a snarl of inextricable confusion. The original *names* of all the units of weights and measures have been improper applications of the substances from which they were derived. Thus, the foot, the palm, the span, the digit, the thumb, and the nail, have been, as measures, improperly so called, for the several parts of the human body, with the length of which they corresponded. Instead of a specific name, the measure usurped that of the standard from which it was taken. Had the foot rule been unalterable, the inconvenience of its improper appellation might have been slight. But, in the lapse of ages, and the revolutions of empires, the foot measure has been everywhere retained, but infinitely varied in its extent. Every nation of modern Europe has a foot measure, no two of which are the same. The English foot indeed was adopted and established in Russia by Peter the Great; but the original Russian foot was not the same. The Hebrew shekel and the maneh, the Greek mina, and the Roman pondo were *weights*. The general name *weight* improperly applied to the specific unit of weight. The Latin word libra, still more improperly, was borrowed from the balance in which it was employed: libra was the balance, and at the same time the pound weight. The terms *weight* and *balance* were thus generic terms, without specific meaning. They signified any weight in the balance, and varied according to the varying gravities of the specific standard unit at different times and in different countries. When, by the debasement of the coins they ceased to be identical with the weights, they still retained their names. The *pound*

sterling retains its name three centuries after it has ceased
to exist as a weight, and after having, as money, lost more
than two-thirds of its substance. We have discarded it
indeed from our vocabulary; but it is still the unit of
moneys of account in England. The *livre* tournois of
France, after still greater degeneracy, continued until the
late revolution, and has only been laid aside for the new
system. The ounce, the drachm, and the grain, are specific names, indefinitely applied as indefinite parts of an
indefinite whole. The English pound avoirdupois is
heavier than the pound troy; but the ounce avoirdupois
is lighter than the ounce troy. The weights and measures
of all the old systems present the perpetual paradox of a
whole not equal to all its parts. Even numbers lose the
definite character which is essential to their nature. A
dozen become sixteen, twenty-eight signify twenty-five,
one hundred and twelve mean a hundred. The indis-
criminate application of the same generic term to different
specific things, and the misapplication of one specific term
to another specific thing, universally pervade all the old
systems, and are the inexhaustible fountains of diversity,
confusion, and fraud. In the vocabulary of the French
system, there is one specific, definite, significant word, to
denote the unit of lineal measure; one for superficial, and
one for solid measure; one for the unit of measures of
capacity, and one for the unit of weights. The word is
exclusively appropriated to the thing, and the thing to the
word. The metre is a definite measure of length : it is
nothing else. It cannot be a measure of one length in one
country, and of another length in another. The gramme
is a specific weight, and the litre a vessel of specific cubic

contents, containing a specific weight of water. The multiples of these units are denoted by prefixing to them syllables derived from the Greek language, significant of their increase in decimal proportions. Thus, ten metres form a deca-metre; ten grammes a deca-gramme; ten litres, a deca-litre. The subdivisions, or decimal fractions of the unit, are equally significant in their denominations, the prefixed syllables being derived from the Latin language. The deci-metre is a tenth part of a metre; the deci-gramme, the tenth part of a gramme; the deci-litre, the tenth part of a litre. Thus, in continued multiplication, the hecto-metre is a hundred, the kilo-metre, a thousand, and the myria-metre ten thousand metres; while, in continued division, the centi-metre is the hundredth, and the milli-metre the thousandth part of the metre. The same prefixed syllables apply equally to the multiples and divisions of the weight, and of all the other measures. Four of the prefixes for multiplication, and three for division, are all that the system requires. These twelve words, with the *franc*, the *decime*, and the *centime*, of the coins, contain the whole system of French metrology, and a complete language of weights, measures, and money.

But where is the steam-engine of moral power to stem the stubborn tide of prejudice, and the headlong current of inveterate usage? The cheerful, ready, and immediate adoption, by the mass of the nation, of these twelve words, would have secured the triumph of the new system of France. The unutterable confusions of signifying the same thing by different words, and different things by the same word, would have ceased. The *setier* would no

longer have been a common representative for twelve boisseaus of corn, for fourteen of oats, for sixteen of salt, and
for thirty-two of coal, and for eight pints of wine. The
pound would no longer have been of ten, of twelve, of
fourteen, of sixteen, and of eighteen ounces, in different
parts of the same country. The weights and the measures
would have been both perfect and just: and the blessing
of uniformity enjoyed by France would have been the
most effective recommendation of her system to all the rest
of mankind. It is mortifying to the philanthropy, which
yearns for the improvement of the condition of man, to
know that this is precisely the part of the system which it
has been found impracticable to carry through.

The modern language of all the mathematical and
physical sciences is derived from the Greek and Latin;
with a partial exception of some terms which are of Arabic
origin. Geography, chemistry, the pure mathematics, botany, mineralogy, zoology, in all of which great discoveries
have been made within the last three centuries, have borrowed from those primitive languages almost invariably
the words by which those discoveries have been expressed.
They are the languages in which all that was heretofore
known of art or science was contained: nor are the moral
and political sciences less indebted to them for numerous
additions to their vocabularies which the progress of modern improvements has required. But there is a natural
aversion in the mass of mankind to the adoption of words,
to which their lips and ears are not from their infancy
accustomed. Hence it is that the use of all technical language is excluded from social conversation, and from all
literary composition suited to general reading; from poetry,

from oratory, from all the regions of imagination, and taste in the world of the human mind. The student of science, in his cabinet, easily familiarizes to his memory, and adopts without repugnance, words indicative of new discoveries or inventions, analogous to the words in the same science already stored in his memory. The artist, at his work, finds no difficulty to receive or use the words appropriate to his own profession. But the general mass of mankind, of every condition, reluct at the use of unaccustomed sounds, and shrink especially from new words of many syllables. But weights and measures are instruments indispensable, not only to the philosophical student and the professional artist; they are the want of every individual and of every day. They are the want of food, of raiment, of shelter, of all the labors and all the pleasures of social existence. Weights and measures, like all the common necessaries of life, have in all the countries of modern Europe, customary names of one, or, at most, of two syllables. The units of the new French system have no more; but their multiples and subdivisions have four or five; and, although compounded of syllables familiar to those who had any acquaintance with the classical languages of Greece and Rome, they had a strange and outlandish sound to the ears of the people in general who would never be taught to pronounce them. Hence, after an experience of several years, it was found necessary, not only to give back to the people the vulgar fractions of their measures, which had been taken from them, but all their indefinite and many-meaning words of pound and ounce, foot, aune, and thumb, boisseau and pint. Since which time there have been, besides all the relics of the old metrology, two concurrent systems of weights and measures in France; one, the proper legal system, with decimal divisions and multiplications, and the new, precise, and significant nomenclature; and the other a system of sufferance, with the same instruments, but divided in all the old varieties of vulgar fractions, and with the old improper vocabulary, made still more so by its adaptation to new and different things.

Perhaps it may be found, by more protracted and multiplied experience, that this is the only uniformity attainable by a system of weights and measures for universal use: that the same material instruments shall be divisible decimally for calculations and accounts; but in any other manner suited to convenience in the shops and markets; that their appropriate legal denominations shall be used for computation, and the trivial names for actual weight, or mensuration.

It results, however, from this review of the present condition of the French system in its native country, and from the comparison of its theoretical advantages over that which we already possess, that the time has not arrived at which so great and hazardous an experiment can be recommended, as that of discarding all our established existing weights and measures, to adopt and legalize those of France in their stead. The single standard, proportional to the circumference of the earth; the singleness of the units for all the various modes of mensuration; the universal application to them of decimal arithmetic; the unbroken chain of connection between all weights, measures, moneys, and coins; and the precise, significant, short, and complete vocabulary of their denominations; altogether forming a system adapted equally to the use of all mankind; afford such a combination of the principle of uniformity for all the most important operations of the intercourse of human society; the establishment of such a system so obviously tends to that great result, the improvement of the physical, moral, and intellectual condition of man upon earth; that there can be neither doubt nor hesitation in the opinion, that the ultimate adoption, and universal though modified application of that system, is a consummation devoutly to be wished.

To despair of human improvement is not more congenial to the judgment of sound philosophy than to the temper of brotherly kindness. Uniformity of weights and measures is, and has been for ages, the common, earnest, and anxious pursuit of France, of Great Britain, and since their independent existence, of the United States. To the attainment of one object, common to them all, they have been proceeding by different means, and with different ultimate ends. France alone has proposed a plan suitable to the ends of all; and has invited co-operation for its construction and establishment. The associated pursuit of great objects of common interest is among the most powerful modern expedients for the improvement of man. The principle is at this time in full operation for the abolition of the African slave-trade. What reason can be assigned, why other objects of common interest to the whole species, should not be in like manner made the subject of common deliberation and concerted effort? To promote the intercourse of nations with each other, the uniformity of their weights and measures is among the most efficacious agencies: and this uniformity can be effected only by mutual understanding and united energy.

A single and universal system can be finally established only by a general convention, to which the principal nations of the world shall be parties, and to which they shall all give their assent. To effect this, would seem to be no difficult achievement. It has one advantage over every plan of moral or political improvement, not excepting the abolition of the slave-trade itself: there neither is, nor can be, any great counteracting *interest* to overcome. The conquest to be obtained is merely over prejudices, usages, and perhaps national jealousies. The whole evil to be subdued is diversity of opinion with regard to the means of attaining the same end. To the formation of the French system, the learning and the genius of other nations did co-operate with those of her native sons. The co-operation of Great Britain was invited; and there is no doubt that of the United States would have been accepted, had it been offered. The French system embraces all the great and important principles of uniformity, which can be applied to weights and measures; but that system is not yet complete. It is susceptible of many modifications and improvements. Considered merely as a labor-saving machine, it is a new power, offered to man, incomparably greater than that which he has acquired by the new agency which he has given to steam. It is in design the greatest *invention* of human ingenuity since that of printing. But, like that, and every other useful and complicated invention, it could not be struck out perfect at a heat. Time and experience have already dictated many improvements of its mechanism; and others may, and undoubtedly will, be found necessary for it hereafter. But all the radical principles of uniformity are in the machine: and the more universally it shall be adopted, the more certain will it be of attaining all the perfection which is within the reach of human power.

Another motive, which would seem to facilitate this concert of nations, is, that it conceals no lurking danger to the independence of any of them. It needs no convocation of sovereigns, armed with military power. It opens no avenue to partial combinations and intrigues. It can mask, under the vizor of virtue, no project of avarice or ambition. It can disguise no private or perverted ends, under the varnish of generous and benevolent aims. It has no final appeal to physical force; no *ultima ratio* of cannon-balls. Its objects are not only pacific in their nature, but can be pursued by no other than peaceable means. Would it not be strange, if, while mankind find it so easy to attain uniformity in the use of every engine adapted to their mutual destruction, they should find it impracticable to agree upon the few and simple but indispensable instruments of all their intercourse of peace and friendship and beneficence—that they should use the same artillery and musketry, and bayonets and swords and lances, for the wholesale trade of human slaughter, and that they should refuse to weigh by the same pound, to measure by the same rule, to drink from the same cup, to use, in fine, the same materials for ministering to the wants and contributing to the enjoyments of one another?

These views are presented as leading to the conclusion, that, as final and universal uniformity of weights and measures is the common desideratum for all civilized nations; as France has formed, and for her own use has established, a system, adapted, by the highest efforts of

human science, ingenuity, and skill, to the common purposes of all; as this system is yet new, imperfect, susceptible of great improvements, and struggling for existence
even in the country which gave it birth; as its universal
establishment would be a universal blessing; and as, if
ever effected, it can only be by consent, and not by force,
in which the energies of opinion must precede those of
legislation; it would be worthy of the dignity of the Congress of the United States to consult the opinions of all
the civilized nations with whom they have a friendly
intercourse; to ascertain, with the utmost attainable accuracy, the existing state of their respective weights and
measures; to take up and pursue, with steady, persevering,
but always temperate and discreet exertions, the idea conceived, and thus far executed, by France, and to co-operate
with her to the final and universal establishment of her
system.

But, although it is respectfully proposed that Congress
should immediately sanction this consultation, and that it
should commence, in the first instance, with Great Britain
and France, it is not expected that it would be attended
with immediate success. Ardent as the pursuit of uniformity has been for ages in England, the idea of extending it beyond the British dominions has hitherto received
but little countenance there. The operation of changes
of opinion there is slow; the aversion to all innovations,
deep. More than two hundred years had elapsed from
the Gregorian reformation of the calendar, before it was
adopted in England. It is to this day still rejected throughout the Russian empire. It is not even intended to propose
the adoption by ourselves of the French metrology for the

present. The reasons have been given for believing, that the time is not yet matured for this reformation. Much less is it supposed advisable to propose its adoption to any other nation. But, in consulting them, it will be proper to let them understand, that the design and motive of opening the communication is, to promote the final establishment of a system of weights and measures, to be common to all civilized nations.

In contemplating so great, but so beneficial a change, as the ultimate object of the proposal now submitted to the consideration of Congress, it is supposed to be most congenial to the end, to attempt no present change whatever in our existing weights and measures; to let the standards remain precisely as they are; and to confine the proceedings of Congress at this time to authorizing the Executive to open these communications with the European nations where we have accredited ministers and agents, and to such declaratory enactments and regulations as may secure a more perfect uniformity in the weights and measures now in use throughout the Union.

The motives for entertaining the opinion, that any change in our system at the present time would be inexpedient, are four:

First, That no change whatever of the system could be adopted, without losing the greatest of all the elements of uniformity, that referring to the persons using the same system. This uniformity we now possess, in common with the whole British nation; the nation with which, of all the nations of the earth, we have the most of that intercourse which requires the constant use of weights and measures. No change is believed possible, other than that of the whole system, the benefit of which would compensate for the loss of this uniformity.

Secondly, That the system, as it exists, has an uniformity of proportion very convenient and useful, which any alteration of it would disturb, and perhaps destroy; the proportion between the avoirdupois and troy weights, and that between the avoirdupois weight and the foot measure; one cubic foot containing of spring-water exactly one thousand ounces avoirdupois, and one pound avoirdupois consisting of exactly seven thousand grains troy.

Thirdly, That the experience of France has proved, that binary, ternary, duodecimal, and sexagesimal divisions, are as necessary to the practical use of weights and measures, as the decimal divisions are convenient for calculations resulting from them; and that no plan for introducing the latter can dispense with the continued use of the former.

Fourthly, That the only *material* improvement, of which the present system is believed to be susceptible, would be the restoration of identity between weights and silver coins; a change, the advantages of which would be very great, but which could not be effected without a corresponding and almost total change in our coinage and moneys of account; a change the more exceptionable, as our monetary system is itself a new, and has hitherto been a successful institution.

Of all the nations of European origin, ours is that which least requires any change in the system of their weights and measures. With the exception of Louisiana, the established system is, and always has been, throughout the Union, the same. Under the feudal system of Europe, combined with the hierarchy of the church of Rome, the people were in servitude, and every chieftain of a village, or owner of a castle, possessed or asserted the attributes of sovereign power. Among the rest, the feudal lords were in the practice of coining money, and fixing their own weights and measures. This is the great source of numberless diversities existing in every part of Europe, proceeding not from the varieties which in a course of ages befell the same system, but from those of diversity of origin. The nations of Europe are, in their origin, all compositions of victorious and vanquished people. Their institutions are compositions of military power and religious opinions. Their doctrines are, that freedom is the grant of the sovereign to the people, and that the sovereign is amenable only to God. These doctrines are not congenial to nations originating in colonial establishments. Colonies carry with them the general laws, opinions, and usages, of the nation from which they emanate, and the prejudices and passions of the age of their emigration. The North American colonies had nothing military in their origin. The first English colonies on this continent were speculations of commerce: they commenced precisely at the period of that struggle in England between liberty and power, which, after long and bloody civil wars, terminated in a compromise between the two conflicting principles. The colonies were founded by that portion of the people, who were arrayed on the side of liberty. They brought with them all the rights, but none of the servitudes, of the parent country. Their constitutions were, indeed, conformably to the spirit of the feudal policy, charters granted by the crown; but they were all adherents to the doctrine, that charters were not donations, but compacts. They brought with them the weights and measures of the law, and not those of any particular district or franchise. The only change which has taken place in England with regard to the legal standards of weights and measures, since the first settlement of the North American colonies, has been the specification of the contents of measures of capacity, by prescribing their dimensions in cubical inches. All the standards at the Exchequer are the same that they were at the first settlement of Jamestown; with the exception of the wine gallon, which is of the time of Queen Anne: and the standards of the Exchequer are the prototypes from which all the weights and measures of the Union are derived.

A particular statement of the regulations of the several States relative to weights and measures, is subjoined to this report, in the Appendix.

The first settlement of the English colonies on the continent of North America was undertaken toward the close of the reign of Queen Elizabeth, in honor of whom it received the name of Virginia.

During the same reign of Elizabeth, and contemporaneous with the adventures which preceded the settlement of Jamestown, the act of parliament of 1592 passed, defining in feet the statute mile. This mile, together with its elementary units, the foot and inch, were the measures by which all the territories, granted by the successors of Elizabeth, in this hemisphere, were defined. The foot and inch, from usage immemorial in England, and by a statute then of more than three centuries' antiquity, had been the elements of superficial, as well as of itinerary land measure. These, therefore, were not only the most natural measures for the use of the English colonies; they were inwoven in their primitive constitutions, and were brought with their charters, an essential part of their possessions.

Among the earliest traces of colonial legistlation in Virginia and in New England, we find acts declaring the assize of London, and the standards of the Exchequer, to be the only lawful prototypes of the weights and measures of the colonies. The foot and inch were of dimensions perfectly well ascertained: and in the year 1601, only seven years before the settlement at Jamestown, and less than twenty before that of Plymouth, new standards, not only of the yard and ell, but of the avoirdupois and troy weights, and of the bushel, corn gallon, quart, and pint, had been deposited at the Exchequer. There was neither uncertainty, nor perceptible diversity, with regard to the long measures or the weights; but the standard vessels of capacity were of various dimensions. The bushel of 1601 contained 2,124 cubic inches; it was therefore a copy from an older standard, made in exact conformity to the rule prescribed in the statute of 1266, and very probably the identical standard therein described. It contained eight corn gallons of wheat, equiponderant to eight Irish gallons of Gascoign wine; of wheat, thirty-two kernels of which were of equal weight with the round, unclipped penny sterling of 1266. Its corresponding wine gallon, therefore, would have been the Irish gallon of 217.6 cubic inches; and its corresponding corn gallon of 265.5 inches, an intermediate between the Rumford quart and gallon of 1228, and differing less than one inch from either of them. There were two other standard bushels at the Exchequer, of the same dimensions; one of the age of Henry the Seventh, and one dated 1091. This has been supposed to be a mistake for 1591 or 1601. But as it is not probable that two standard bushels should have been deposited in the Exchequer at the same time, or even at dates so near to each other, a conjecture may be indulged, that the 1091 marks the date, when the standard measure, described in the statute of 1266, was made. Of that standard, these three bushels were unquestionably copies.

The corn and the ale gallons of 1601 were of 272 cubic inches; and there was one of Henry the Seventh there, of the same size, as reported by the artist who measured them for the commissioners of the excise in 1688. When measured again by order of the committee of the House of Commons, in 1758, they were reported to contain each about one inch less. The true size intended for all of them was 272; and they were made by an application of the rule of 1266 to the troy weight wheat of the act of 1496. They were the eighth parts of a bushel of 2,176 inches; and their corresponding wine gallon was the Guildhall gallon of 224 inches.

There were, in 1601, a standard quart of 70 inches, and a pint of 34.8; which were evidently intended to be in exact proportions to each other: and the gallon, to which they referred, was the gallon of 282 inches. This would have made a bushel of 2,256 inches; and its corresponding wine gallon is of 231 inches. The standards, thus made, were by an application both of the wheat and of the rule described in the statute of 1266 to the troy weight gallon of 1496; that is, the wheat was of the kind, 32 kernels of which weighed the same as the old penny sterling, and of which the wine gallon contained eight pounds troy weight. There was a standard bushel of Henry the Seventh at the Exchequer, of 2,224 inches, probably the bushel from which this quart and these pints were deduced.

There was also the *Winchester* bushel of 2,145.6 cubic inches, made in the reign of Henry the Seventh, but from its name evidently copied from a standard which had been kept at Winchester when that place was the capital of the kingdom. This bushel had been made, by combining the rule of 1266 with the assize of casks which, in the statute of 1423, is declared to be of *old time*, by which the hogshead, or eight cubic feet of Gascoign wine, consisted of 63 gallons. That hogshead was a quarter of a ton of wine, as eight Winchester bushels contained a quarter of a ton of wheat. The gallon was of 219½ cubic inches; and the corresponding ale gallon was of 268.2 inches. There was at the Exchequer no wine or ale gallon of those dimensions; because the wine gallon of 224 inches, and the corn gallon of 272, made under the statutes of 1496 and 1531, had been substituted in their stead. At the Exchequer, there was indeed no wine gallon at all. Those of older date than the act of 1406 had disappeared, and the gallon of 224 inches made according to that act, had been delivered out of the Exchequer to the city of London, and was at Guildhall.

Such was the state of the standards in London, at the time of the first colonial emigrations to this continent.

MASSACHUSETTS.

Among the colony laws of Massachusetts, there is an act of the year 1647, directing the country treasurer *to provide,* at the country's charge, weights and measures *of all sorts* for continual standards. In the specification which ensues in the act, all the measures, of which there were standards at the Exchequer, are mentioned, with special discrimination of *wine* and *ale measures;* but the weights only *after sixteen ounces* to the pound, are named. They then had no occasion for the troy weights.

At a still earlier date, in 1641, it had been prescribed that all casks for any liquor, fish, beef, pork, or other commodities to be put to sale, should be *of London assize:* and in 1646 a corresponding assize of *staves* had been ordained.

The law of 1647 did not expressly direct where the treasurer was to procure the standards : but the Exchequer and Guildhall were the only places where they were to be obtained ; and from subsequent acts, the fact appears that they were obtained there.

At the first session of the general court under the charter of William and Mary, in 1692, two laws were enacted ; one, reordaining the London assize of casks, and specifying that the butt should contain 126 gallons, the puncheon 84, the hogshead 63, the tierce 42, and the barrel 31½ gallons ; the other, for due regulation of weights and measures, declaring that the brass and copper measures, *formerly sent* out of England, with certificate out of the Exchequer to be approved *Winchester* measure, according to the standard in the Exchequer, should be the public allowed standard throughout the province for the proving and sealing all weights and measures thereby, and re-enacting, with an additional clause, the colonial law of 1647.

An act of the year 1700 prescribes, that the bushel used for the sale of meal, fruits, and other things, usually sold by heap, shall be not less than 18½ inches wide within side; the half-bushel not less than 13¾ inches; the peck not less than 10¾, and the half-peck not less than 9 inches.

It is very remarkable that this law was enacted one year *before* the act of parliament of 13 William III. which gives and prescribes in cubical inches the dimensions of the Winchester bushel. The object of the provincial law was, to prohibit the use of bushels, which, though of the same cubical capacity, should be of shorter diameter and greater depth. It was for the benefit of the heap. It prescribed, therefore, only the diameter, without mentioning the depth; but that diameter, for the bushel, is identically the same, 18½ inches, as the act of parliament of the ensuing year declares to be the width of the Winchester bushel in the Exchequer. As the provincial standard must have been the model from which the law of the province took its measure of a diameter, its perfect coincidence with the subsequent definition of the act of parliament, is a proof of the correctness of the copy from the Winchester bushel of the Exchequer.

In 1765 the treasurer of the province was required by law to procure a beam, scale, and a nest of *troy weights* from 128 ounces down, marked with a mark or stamp used at the Exchequer, for a public standard. Every town was to be provided with a nest of troy weights of different form from the avoirdupois, and a penalty was annexed to the use of any other than sealed troy weights for weighing silver, bullion, or other species whatsoever, proper and *used* to be weighed by troy weights.

In the year 1707 there was an act of parliament, 6 Anne, ch. 30, " for ascertaining the rates of foreign coins in her majesty's plantations in America." It had been preceded, in 1704, by a proclamation of the queen declaring the value of many foreign silver coins, and particularly of the Spanish piece of eight, or dollar. At that period, as in a certain degree at the present, the Spanish dollar and its parts formed the principal circulating coins of this country. The act declares the value of the Seville, Pillar, and Mexican pieces of eight, to be four shillings and six-pence sterling, and their weight to be seventeen pennyweights and a half, or 420 grains. It forbids their being taken in the colonies at more than six shillings each, and this act constituted what, from that time till the period of the Revolution, in Virginia and New England, was denominated " lawful money." The act itself was published in the province of Massachusetts Bay with the statutes of the provincial legislature, as was practised with regard to all the acts of parliament, the authority of which was recognized. It may be here incidentally remarked, that the laws of Congress, which estimate the value of the English pound sterling at four dollars and forty-four cents, are all founded upon the proportions established by this act, although the weight and value, both of the dollar and of the shilling and pound sterling, have since that time been changed. Some further observations on this subject are submitted in the Appendix, from which it will appear that the real value of our silver dollar, in the silver English half-crowns or shillings of this time, is four shillings, seven pence, and nearly one farthing; and that the pound sterling of such actual English silver coins is, in the silver money of the United States, not 4 dollars 44 cents, but only 4 dollars, 34 cents, and 9 mills.

In the year 1715 a lighthouse was built at the entrance of Boston harbor; and a tonnage duty being levied upon vessels entering the harbor, to defray the expense of building and supporting it, the rule of measurement for ascertaining the tonnage prescribed by the act was, that a vessel of two decks should be measured, upon the main deck, from the stem to the stern-post, then subducting the breadth, from outside to outside, athwart the main beam, the remainder to be accounted her length by the keel, which, being multiplied by the breadth, and the product by one-half the breadth for the depth, and the whole product divided by 100, the quotient was to be accounted the tonnage of the ship. Vessels of a single deck, or $1\frac{1}{2}$ deck, were to be measured in the same manner, except the depth in hold, which was to be from the under side of the main beam to the ceiling.

In 1730 a new set of brass and copper avoirdupois weights and of measures, was imported from the Exchequer, with certificate of their being approved Winchester measure, according to the standard in the Exchequer. These were, by a new statute, declared to be the public standards of the province: and they continue to be those of the commonwealth at this day. It does not appear that the troy weights were renewed at the same time. The standards of them had been imported only twenty-five years before, and could not need renewing.

In 1751 the act of parliament introducing the Gregorian calendar was adopted, in the usual manner, by inserting it among the laws of the provincial legislature.

Since the Revolution, all the laws of the province have
been revised: and by an act of the legislature of 26th
February, 1800, all the principal regulations concerning
weights and measures were renewed and confirmed.

This law declares that the brass and copper measures
formerly (1730) sent out of England with a certificate
from the Exchequer, shall be and remain the public standards throughout the commonwealth: and it requires the
treasurer of the commonwealth to cause to be had and preserved a complete set of new beams, weights, avoirdupois
and troy, and measures of length and of capacity, wet and
dry, to be used only as public standards. This act is to
continue until Congress shall have fixed by law the standard of weights and measures.

A statute of 9th March, 1804, recites, that the troy
weights used by the treasurer of the commonwealth, as State
standards, had, by long use, diminished and undergone
an alteration in their proportions. (They had then been
just one century in use.) It directs him, therefore, to add,
or cause to be added, a specified number of grains to each
of the weights, from that of 128 ounces to the half-ounce;
or to procure new weights of the same denomination, and
conformable to the State standards, with such additions,
which weights, so corrected, are declared to be the standards of troy weight for the commonwealth. By information from various sources, it is known that the standards
of the State of Massachusetts are, at this time, perfectly
conformable to those of the Exchequer.

There are a multitude of laws regulating the assize of
casks, assigning different dimensions for containing different articles. They generally prescribe the length of the

staves within the chime, and sometimes the diameter of the heads. They also specify the weight of the article which the cask is to contain. Staves are an article of exportation, and their length, breadth, and thickness, are regulated by law.

NEW HAMPSHIRE AND VERMONT.

The laws of New Hampshire, and of Vermont, relating to weights and measures, appear to have been modelled upon those of Massachusetts. In both these States the standards are required to be according to the approved Winchester measures, allowed in England, in the Exchequer. The first act of New Hampshire to that effect was of 13th May, 1718, and the last of 15th December, 1797. The statute of Vermont is of the 8th of March, 1797. Neither New Hampshire nor Vermont has established the authority of the troy weights by law.

RHODE ISLAND.

Rhode Island has no statute upon the subject. Her weights and measures are, however, the same, and her standards are taken from those of Massachusetts.

CONNECTICUT.

In the laws and standards of Connecticut there are peculiarities deserving of remark.

A statute of October, 1800, contains the following provisions: "That the brass measures, the property of this State, kept at the Treasury, that is to say, a half-bushel measure, containing *one thousand and ninety-nine* cubic inches, very near, a peck measure and half-peck measure, when reduced to a just proportion, be the standard of the corn measures in this State, which are called by those names respectively; that the brass vessels ordered to be provided by this Assembly [one of the capacity of two hundred and twenty-four cubic inches]* and the other of the capacity of two hundred and eighty-two cubic inches, shall be, when procured, the first of them the standard of a wine gallon, and the other the standard of an ale or beer gallon, in this State; that the iron, or brass rod or plate, ordered by this Assembly to be provided, of one yard in length, to be divided into three equal parts, for feet, in length, and one of those parts to be subdivided into twelve equal parts, for inches, shall be the standard of those measures respectively; and that the brass weights, the property of the State, kept at the Treasury, of one, two, four, seven, fourteen, twenty-eight and fifty-six pounds, shall be the standard of avoirdupois weight in this State."

A subsequent section (5) requires of the selectmen of each town to provide town standards, of good and suffi- cient materials, which, for the standards of liquid measure, shall be copper, brass, or pewter; also, vessels for corn measure, of forms and dimensions thus described: "A two- quart measure, the bottom of which, on the inside, is four inches wide on two opposite sides, and four inches and a half on the two other sides, and its height from thence seven inches and sixty-three hundredths of an inch;" [137.34 cubic inches]. A quart measure of "three inches square from bottom to top, throughout, and its height

---

* These brackets, in the printed volume of the laws of Connecti- cut, indicate that the part enclosed has been repealed.

seven inches and sixty-three hundredths of an inch;" [68.67 cubic inches]. A pint measure of three inches square from bottom to top throughout, and its height three inches and eighty-two hundredths of an inch; [34.38 cubic inches].

The assumption of the old Guildhall wine gallon, of 224 inches, in this act, is the more surprising, inasmuch as a colonial statute of the year 1752 had already established the gallon of 231 inches. What the occasion of it was, has not been ascertained; but it was probably taken from an existing standard, which had been originally taken from the Guildhall gallon. Whatever the cause of it may have been, this part of the act was repealed the next year (October, 1801), and the Treasurer was directed, without delay, to provide a vessel of brass, of five inches square from bottom to top throughout, and nine inches and twenty-four hundredths of an inch in height, containing two hundred and thirty-one cubic inches, which was declared the standard wine gallon of the State.

The half-bushel measure, which in 1800 was the property of the State, kept at the Treasury, containing 1099 cubic inches, very near, was of course not originally derived from the Winchester bushel. By the colonial laws of Connecticut it appears that, as early as the year 1670, there were *colony standards* kept at Hartford: and the half-bushel, which in the year 1800 was there at the Treasury, the property of the State, was either one of those same standards of 1670, or a copy from it. That it was not borrowed from the Massachusetts standards is also manifest, because the Massachusetts bushel was copied from the Winchester bushel. It may be concluded, with great probability, that the Connecticut half-bushel was first taken from the bushel in the Exchequer of Henry the Seventh, with a copper rim; though it contains thirteen cubic inches less than in proportion to that standard. This difference, in so large a measure, may have been the effect of very slight inaccuracy in the first copy, increased by the decay or the change of the vessel. The bushel with the copper rim was deposited at the Exchequer after the act of 1496; and was made from the wine gallon of that act, with the rule of the act of 1266, and the pound of fifteen ounces troy weight. The quart and pint at the Exchequer of 1601, were formed from this bushel. The pint differs less than half an inch from that prescribed by this act of Connecticut of 1800.

In the laws of Connecticut, as in those of New Hampshire and Vermont, there is no formal establishment or recognition of troy weights; nor does there appear to be any standard of them existing in the State. But in the lists of rateable estate, prescribed by the laws of Connecticut, silver plate is estimated at one dollar eleven cents per ounce, which must obviously be intended the ounce troy.

The assize of casks is regulated by various laws: and the dimensions of the barrel for packing salted provisions for exportation are the same as those established in Massachusetts and New York. The London assize of *tight* casks, from the puncheon of 126 to the barrel of 31½ gallons, was coeval with the first legislation of the colony; and was re-enacted by a statute of 1795. It expressly declares that these gallons shall be of 231 cubic inches; and directs that they shall be computed by taking, in inches and decimal parts of an inch, the bulge or bung diameter, each head diameter, and the length within the cask, with Gunter's rule of gauging.

The assize of staves is the same as in Massachusetts.

By an act of October, 1796, the standard weight of wheat is declared to be sixty pounds net to the bushel.

NEW YORK.

New York was originally the seat of a colony from the Netherlands, the settlers of which doubtless brought with them the weights and measures of their own country. Toward the close of the seventeenth century it fell into the possession of the English; and on the 19th of June, 1703, an act of the colonial legislature established all the English weights and measures, *according to the standards in the Exchequer*. This act was drawn with great care, and evidently with the purpose of embracing all the provisions of the then existing English statutes, regulating weights, measures, and casks, particularly those of 1266, 1304, 1439, and 1496, without being aware of the utter incompatibility of those statutes with one another.

Instead, however, of adopting in terms the London assize of casks, from the ton of 252 gallons downward, this act prescribes in inches the length and head diameters of the various casks; and, by a very remarkable peculiarity, changes the names of all the dry casks. It directs that

The Hogshead shall be 40 inches long, 33 inches in the bulge, 27 in the head.
 Tierce " " 36 " " 27 " " " " 23
 Barrel " " 30 " " 26 " " " " 22
 Half-barrel " " 25 " " 20 " " " " 16
 Quarter-barrel " 20 " " 16 " " " " 13

But it adds, that *tite* barrels shall contain 31½ gallons wine measure, or within a gallon more or less, and all other casks in proportion. This last provision adopted the whole London assize for tight casks. But the dimensions prescribed for the *hogshead*, give a cask of about 126 gallons, which, in the London assize, made the butt or pipe; and thus the New York tierce was of 80 gallons, which constituted the *real* contents of the London puncheon; the New York barrel was of 60 gallons, answering to the London hogshead; and the New York half-barrel of 30 gallons, to the London barrel.

On the 10th of April, 1784, the legislature of New York passed an act to *ascertain* weights and measures within the State. It declares the standard weights and measures which were in the custody of William Hardenbrook, public scaler and marker in the city and county of New York, at the time of the declaration of Independence, which were according to the standard of the Exchequer, to be the standard throughout the State. William Hardenbrook was directed to deliver them to the clerk of the city and county of New York, and to make oath that they were the same which he had received from the Court of Exchequer.

By an act of 7th March, 1788, the standard weight of wheat brought to the city of New York for sale, was fixed at sixty pounds net to the bushel.

On the 24th of March, 1809, passed an act relative to a standard of long measure, and for other purposes. It declares a brass yard-measure, engraved and sealed at the Exchequer of Great Britain, procured in 1803 by the corporation of New York, presented to the State, and deposited, with authenticating documents, in the secretary's office, to be the standard yard-measure of the State.

The last statute, upon this subject, of New York, is an act, to regulate weights and measures, and passed on the 19th of March, 1813; which declares that there shall be one just beam, one certain weight and measure for distance and capacity; that is to say, avoirdupois and troy weights, bushels, half-bushels, pecks, half-pecks, and quarts; and gallons, half-gallons, quarts, pints, and gills; and one certain rod for long measure, according to " the standard in use in the State on the day of the declaration of the Independence thereof, and that the standard of weights and measures in the office of the Secretary of the State, which is according to the standard in the Court of Exchequer in that part of Great Britain called England, shall be and remain the standard for ascertaining all beams, weights, and measures throughout the State, until the Congress of the United States shall establish the standard of weights and measures for the United States."

The assize of casks continues as it was regulated by the act of 1703: but a variety of special statutes assign dimensions different from it for barrels in which beef, pork, fish, flour, pot and pearl ashes, etc., are packed for exportation. These, as in New England States, are adapted to contain a certain specified weight of each article. The assize of *staves* regulated by an act of 26th March, 1813, is substantially the same as that of Massachusetts: and as the capacity of the barrel must always depend in a great degree upon the size of the staves and heading of which it is made, the contents of all these barrels vary little from 30 gallons wine measure of 231 cubic inches equal to 6,930 inches.

NEW JERSEY.

In New Jersey, which was originally a part of the Dutch settlement, the English weights and measures were established at a later period than in New York. An act of the colonial legislature, of 13th August, 1725, recites, in its preamble, that nothing is more agreeable to common justice and equity than that throughout the province there should be *one* just weight and balance, one true and perfect standard for measures, for *want* whereof experience had shown that many frauds and deceits had happened; for remedy of which, it establishes, in the first section, an assize of casks for packing of beef and pork, since altered; and in the second, declares, that there shall be one just beam and balance, one certain standard for "weights, that is to say: for avoirdupois and troy weights, one standard for measures, bushels, half-bushels, pecks, and half-pecks, one just standard for liquid measures, that is to say: wine and beer measure; and one yard; all which shall be according to the standard of the Exchequer in Great Britain."

The phraseology of this statute has some resemblance to that of the 25th chapter of Magna Charter, and may serve as a lucid commentary upon it; for, although its avowed object is uniformity, and even *unity* of standard, it expressly sanctions two weights, avoirdupois and troy, and two liquid measures for wine and beer. This statute, also, as well as that of New York of 1813, shows that the term *gallon* is improperly used when applied to dry measure, its real denomination being that of half-peck.

The laws of New Jersey relating to the assize of barrels have been various. By an act of 1774, revived in 1783, the barrel is required to contain 31½ wine gallons, and not ½ a gallon more or less ; half-barrels 16 gallons, and not one quart more or less. The assize of staves (26th September, 1772) is materially the same as in all the eastern States.

PENNSYLVANIA.

In the year 1700, two laws relating to weights and measures were enacted by the colonial legislature. The first [Laws of Pennsylvania, Bioren's edition, vol. 1, page 18] ordains, that brass standards of weights and measures, according to the standards for the Exchequer, should be *obtained*, and kept in each county. Sec. 2. That a brass half-bushel then in Philadelphia, and a bushel and peck proportionable, and all lesser measures and weights coming from England, being duly sealed in London, or other measures agreeable therewith, should be accounted good till the standard should be obtained. Sec. 3. That no person should sell beer or ale by retail, *but by beer measure*, according to the standard of England.

The second, not only adopted the London assize of casks, but required that *all* tight casks, for beer, ale, cider, pork, beef, and oil, and all such commodities, should be made of good, sound, well-seasoned white-oak timber, and contain,

The Puncheon	-	-	-	84 gallons
Hogshead	-	-	-	63 "
Tierce	-	-	-	42 "
Barrel	-	-	-	31½ "
Half-barrel	-	-	-	16 "

wine measure, according to the practice of the neighboring colonies.

This act regulated the assize of staves for hogsheads and

barrels; and prescribed that tobacco hogsheads should be four feet long, or within an inch more or less, 32 inches in the head, equal to the gauge of Maryland, and be four hogsheads to a ton; that the flour cask should be not above *double* the gauge of wine measure; the half-barrel to be of 31½ gallons, and the barrel of 63 gallons wine measure.

As the gauge of Maryland was adopted for tobacco, so that of New York was assumed for flour, by constituting the barrel and half-barrel at double the gauge of wine measure. The origin of this must have been in the measures of the Dutch colonies, which had reference to the *last*, or double *ton* of shipping, the customary measure of the Netherlands, instead of the ton.

But the most remarkable peculiarity of these two laws of Pennsylvania, enacted at the same session of the legislature, was, that while one of them applied the London assize of wine measure to the casks which were to contain *beer*, ale, and cider, the other expressly prohibited the retailers of beer and ale from selling those liquors otherwise than by beer measure: so that the retailers were obliged to buy by the small and to sell by the large measure. This inconsistency between the two statutes will not surprise us when we recollect that it occurred precisely at the time when the trial in the Court of Exchequer of England was litigated, concerning the duties to be paid on Mr. Thomas Barker's importation of Alicant wine. For while he, upon a claim to pay duties upon wine only by beer measure, was reducing the Attorney-General, after a trial of five hours, to withdraw a juror and cast the remedy upon parliament, the legislature of Pennsylvania, by the same erroneous application of the same name to different things, were, certainly without intention, but, in effect, enjoining upon all the publicans of the province to pay for beer by wine measure. It was a whimsical operation of the same incongruity happening in the two hemispheres at the same time, that, while Barker was struggling successfully against the supreme authority of the mother country, to pay for wine by beer measure, the Pennsylvania publicans, by the acts of their provincial legislature, were compelled to pay for beer by wine measure, and yet to be paid for it by its own.

The remedy to these disorders was applied in England and in Pennsylvania also about the same time. In both cases, however, it was partial; applied only to the special inconvenience without reaching the source of the evil. Parliament only defined the capacity of the wine gallon, fixing it at 231 cubic inches. The Pennsylvania legislature, by an act of 1705 [P. L. Bioren's edition, vol. 1, ch. 138, p. 43], reciting the inconsistent provisions of their two acts of 1700, and ingenuously remarking, that in consequence of them, *retailers are obliged to sell by far greater measure than they buy*, released them from this burdensome obligation, by authorizing innkeepers to sell beer by wine measure in their houses, and by beer measure to persons to carry it out of the house. The real evil, in both cases, had proceeded from calling the two different measures of liquids by the same name. If the beer gallon had been called a *half-peck*, no such questions, and no such clashing legislation, would ever have arisen. The statute of 1700, which had prescribed the London assize of casks, was repealed only in March, 1810.

The assize of staves and headings was fixed, in Pennsylvania, by a statute of 1769 [chap. 439, vol. 1, p. 222]. It was, with slight variations, the same as in all the States eastward of it. The necessary width of all staves, for exportation, was, by this act, fixed at 3½ inches. By a subsequent act [30th March, 1803, ch. 2,362, vol. 4, p. 83], staves of three inches wide are allowed as merchantable. Uninspected staves or heading may, by an act of 1796 [ch. 1501, vol. 2, p. 529], be used within the State. A great multitude of statutes in Pennsylvania, as in all the other navigating States, have regulated the assize of casks, adapting them to contain weight of the respective articles to be exported in them, and to the convenience of stowage in ships. This, as has been shown, was the original foundation of the London assize of the ton, and of the whole English system of weights and measures: and this, in the act of Pennsylvania of 12th September, 1789 [ch. 1442, vol. 2, p. 490], is expressly assigned as one of the reasons for requiring casks of given dimensions.

DELAWARE.

In 1705, "An act for regulating weights and measures," directs that each county should obtain standard brass weights and measures, according *to the queen's standards* for the Exchequer; that a standard brass half-bushel should be taken from that in Philadelphia, to which the bushel and peck should be proportionable. It authorizes the use of measures and weights coming from England, duly stamped in London, or others agreeable therewith, till the standards should be procured: and it prescribes that beer and ale should be sold in retail only by beer measure.

Subsequent acts of the legislature of Delaware define the cord of fire-wood, rate gold and silver coins by their weight in troy pennyweights and grains; and regulate the assize of casks for flour, corn, and Indian meal, in exact conformity to that of Pennsylvania.

MARYLAND.

The first act concerning weights and measures to be found in the printed editions of the statutes of this State is of the year 1715, ch. 10. "An act relating to the standard of English weights and measures," the preamble of which, recites that the standards are very much impaired in several of the counties of the province, and in some wholly lost or unfit for use. It therefore directs the justices of the several county courts to cause the standards they already had to be made complete, and to purchase new standards where they had none: and requires them to take security from the standard-keepers for the due execution of their office, and the safe-keeping of the standards in future.

What the standards were, is ascertained by recurrence to the records of the State for the laws, the titles only of which are given in the printed compilations of the statutes.

In 1637, at the first general assembly of which any record is extant, *a bill for corn measures* is one of forty-two which were prepared and propounded to the lord proprietary for his assent, but which were not enacted into laws, nor is there any copy of them to be found upon the record.

The next year, 1638, an act for measures and weights

was one of thirty-six bills twice read and engrossed, but never read a third time, nor passed the House. There were in this bill several remarkable peculiarities. It provided that there should be one standard measure throughout the province, to be appointed by the lieutenant-general, and a sealer of measures: that all contracts made for the payment of corn should be understood of corn shelled; that a barrel of new corn, tended in payment at or afore the 15th of October, in any year, should be twice shaked in the barrel, and afterward heaped as long as it will lye on; and at or before the feast of the nativity, should be twice shaked and filled to the edge of the barrel, or else not shaked, and heaped as before; and after the said feast it should not be shaken at all, but delivered by strike. No steelyards or other weights not sealed by the lieutenant-general, or by the sealer appointed by him, were to be used, except it be small weights *sealed in England*. The act was to continue till the end of the next general assembly.

In 1641 there passed *an act for measures*, which, after reciting the inconveniences from the want of a set and appointed measure, whereby corn and other grain might be bought and sold within the province, provides that from thenceforth *the measure used in England, called the Winchester bushel*, should be only used as the rule to measure all things sold by the bushel or barrel; and the barrel was to contain five such bushels. The sheriff of each county was to procure and keep such a standard bushel, whereby others should be sized and sealed, and penalties were affixed to the use of any others.

This act was to continue only two years, and then expired; but *the Winchester bushel* has, from the time of its enactment, remained the standard dry measure of Maryland.

In 1671 passed an act for providing a standard, with English weights and measures, in the several and respective counties within this province. And this statute, though omitted in all the late printed editions of the laws of Maryland, established the standard recognized by the existing act of 1715, and by all the subsequent laws of Maryland relating to the subject.

The preamble complains, that much fraud and deceit is practised in the province, by false weights and measures: for prevention of which it enacts—

That no inhabitant, or *trader hither*, shall use in trading any other *weights* or measures than are used and made, *according to the statute of Henry the Seventh, King of England*, in that case made and provided [the statute of 1496].

That, for the discovery of abuses, nine persons, who are indicated by name, one for each county then in the province, should set up a standard at their own houses, and provide by the next shipping, or the shipping then next following at farthest, twelve half-hundred weights, a quartern, half-quartern, seven pounds, four pounds, two pounds, and one pound; also, each person six stamps for making stillyards and weights, to be lettered from A to I, one letter for each county; also, each person to have nine irons, numbered from one to nine, and another with cypher, for the numbering of stillyards and pea, that they might not be changed, and to procure brass measures of ell and yard, to be sealed in England; also, a sealed bushel, half-bushel, peck, and gallon, of Winchester measure, and gallon, pottle, quart, pint, and half-pint, of wine measures, with three burnt stamps for the wooden measures and three other stamps for the pewter measures, to be all of the same letter with their other stamps; and that these weights, measures, and stamps, should be kept by those nine persons at their respective houses, to which all persons were to bring their stillyards to be tried, stamped, and numbered, once a year, and also their barrels, which were to contain five bushels, and other measures, to be sealed.

The act further provides penalties for using other weights and measures, and, in case of the death of any of the nine persons named as standard-keepers, directs that other persons should be appointed by the commissioners of the respective county courts in their stead.

The limitation of the act was to three years, or the end of the next general assembly. It was revived and continued by several successive acts till 1692, when there passed "An act for the *settling of a standard* with English weights and measures within the several and respective counties in this province."

This is in substance a re-enactment and confirmation of the statute of 1671, providing that the justices of the county courts should, from time to time, appoint a person in each county to keep the standards, and to provide all such weights and measures as were wanting, according to the directions of the act of 1671, and an additional set for Cecil county, with stamps to be marked K.

1704, September 21, ch. 71, An act *relating* to the standard of English weights and measures, has the following preamble:

" Whereas *there is now* a standard of weights and measures *agreeable to the standard of weights and measures in her majesty's Exchequer* in England settled within the several counties of this province."

After this preamble, the act directs, that all persons, whether inhabitants or foreigners, shall bring their still-yards, with which they weigh and receive their tobacco, every year to be tried, stamped, and numbered; and every person, trading with bushels, half-bushels, etc., shall have them tried and stamped at the standard, except such as come out of England and are there stamped: and penalties are prescribed for buying or selling by stillyards or dry measures not thus tried and stamped, but they are not extended either to the weights or the liquid measures.

The titles only of all these statutes are given in the printed editions of the statutes of Maryland. But the parts of them which prescribe the standard are yet in full force. The law, is the memorable act of parliament of 1496: and the fact, in Maryland as in England, is, that the standards have been copied from those in the Exchequer.

In 1765 (1st Nov., ch. 1) was passed a supplementary act to the act of 1715, already noticed, entitled "An act relating to the standard of English weights and measures."

The preamble recites, that, in the act of 1715, there is no penalty upon *buyers* by unstamped dry measures, as there is upon sellers; whence persons refuse to *buy* grain, flaxseed, and other commodities, unless by measures larger than the standard.

It therefore prohibits, upon £5 penalty, buying by such measures.

Neither of these two acts takes any notice either of

long or liquid measures, or of weights. But the standards
had been established by the statute of 1671, and have continued to this time. Beer measure appears never to have
been formally established by the statute law of Maryland:
but troy weight is explicitly recognized in the act of November, 1781 (ch. 16), to declare what foreign gold and
silver coin shall be deemed the current money of the
State. It fixes the value of several of those coins, proportionable to their weight, in ounces, pennyweights, and
grains, intending, though not naming, troy weight; but
rating Spanish milled pieces of eight at seven shillings and
six pence, and French and English crowns at eight shillings and four pence.

In 1796, by an act to erect Baltimore, in Baltimore
county, into a city, and to incorporate the inhabitants
thereof, the corporation (sec. 9) are empowered to regulate
and fix the assize of bread; to provide for the safe-keeping
and preservation of the standard of weights and measures
used within the city and precincts; also, to regulate the assize of bricks, etc. And, in 1805, by an act supplementary
to the act of incorporating Baltimore as a city, it is ordained, Congress not having yet fixed any standard of
weights and measures, that the mayor and city council
shall have and exercise the right of regulating all weights
and measures within the city and precincts by the present
standard, until one shall be determined on by Congress.

The assize of casks has been in Maryland, as in the other
parts of the Union, both before and since our Revolution,
a subject of frequent and voluminous legislation. As early
as the year 1658, there had passed an act, concerning the
gauge of tobacco hogsheads, which had prescribed the

length and diameter at the head of those casks, the dimensions of which were then the same as those used in Virginia. In 1676, this law was re-enacted with some additional sections, and was from time to time continued until 1732.

In November, 1763, by an act for amending the staple of tobacco, etc., the hogsheads containing that article were required to be 48 inches in the length of the stave, and 70 inches in the whole diameter within the staves, at the croze and bulge; a regulation repeated in the act of November, 1801, to regulate the inspection of tobacco, which is now in force.

In 1745, there passed an act for the gauge of barrels for pork, beef, pitch, tar, turpentine, and tare of barrels for flour or bread. It did not prescribe the dimensions of flour and bread casks; but directed that all barrels, made or used for either of those articles, should be of the size and gauge to contain at least the quantity of $31\frac{1}{2}$ gallons wine measure, and that the contents of every pork or beef barrel, for exportation or sale, should be at least 220 pounds net of meat.

This act, though originally limited in duration to three years, and the end of the next session of the assembly, has, by successive re-enactments always limited, been continued in force to this day.

Another act, of 1786, for the inspection of salted provisions, exported and imported from and to the town of Baltimore, required that the staves of beef and pork barrels should be 29 inches long, and 18 inches diameter at the head. And these regulations, though superseded at Baltimore by the exercise of the powers vested in the corporation of that city, have been extended to other parts of the State, and are yet in force.

The size of fish barrels had been prescribed by the same act. But, in February, 1818, by an act to regulate the inspection of salted fish, it was directed, that the barrel staves should be 28 inches in length, the heads 17 inches between the chimes, and to contain not less than 29, nor more than 31 gallons; tierces to hold not less than 45, and half-barrels not less than 15 gallons.

VIRGINIA.

Among the earliest records of the general assembly of the colony of Virginia, is an order of the 5th of March, 1623–4, that there be no weights nor measures used, but such as should be sealed by officers appointed for that purpose.

By an act of 23d February, 1631–2, it was ordained, that a barrel of corn should be accounted five bushels of *Winchester measure*, 40 gallons to the barrel. The commissioners of the monthly courts were to keep sealed barrels, and to seal such as should be brought to them. Whoever used unsealed barrels or bushels was to forfeit thirteen shillings and four pence, and sit on the pillory; and the measure and barrel deficient was to be broken and burnt. And for defective *weights*, it was ordained that the offender should be punished according to the statute in that case provided.

An act of 5th October, 1646, declares, that merchants and others, as well *Dutch* as English, practise deceit by *diversity* of weights and measures used by them; and enacts, that no merchant or trader, whether English or

Dutch, shall trade with other weights and measures, than according to the statute of parliament in such cases provided. What this statute of parliament was, is explained by an act of 23d March, 1661-2, which declares, that, "Whereas dayly experience sheweth that much fraud and deceit is practised in this colony by false weights and measures," for prevention thereof, no inhabitant, or trader hither, shall trade with any other weights or measures than are used and made according to the statute of 12 Henry VII., ch. 5 [the statute of 1496], in that case provided; and that, for discovery of abuses, county commissioners shall provide sealed *weights* of half hundreds, quarterns, half quarterns, seaven pounds, fower pounds, two pounds, one pound, measures of ell and yard, of bushel, half bushel, peck, and gallon, of Winchester measure; gallon, pottle, quart, pint, half pint, of wine measure out of England; to be kept by the first of every commission at the house, and a burnt mark of (cv.) and a stamp for leaden weights and pewter potts, whither all persons, not using weights and measures brought out of England, and sealed there, shall bring all their barrels (which are to contain five bushels) and other measures to be sealed and their stillyards to be tried. Then follow penalties (in tobacco) for selling by other than sealed weights and measures, and upon commissioners for not providing standards.

Thus in Virginia, as in Maryland, the English statute of Henry VII. of 1495, has for near a century and a half been nominally the law of the land concerning weights and measures; while, at the same time, the actual weights and measures of capacity have been copies from the standards in the Exchequer, not one of which has ever been conformable to the statute of 1496. And this very act of Virginia, of 1661, while establishing by law the exclusive *troy* weight, *wine* gallons, and never-made bushel, of the English act of 1496, requires of the county commissioners to provide the *avoirdupois* weights and the *Winchester* measures of the English Exchequer

In the year 1734, a new and amendatory act " for more effectual obliging persons to buy and sell by weights and measures according to the English standard," repeated all the principal provisions of the act of 1661, omitting, however, all reference to the English act of parliament of 1496. And since the Revolution, by an act of the legislature of Virginia, of 26th December, 1792, this act of 1734 is continued, to remain in full force until the Congress of the United States shall have otherwise provided.

Among the numerous wise and honorable examples, which the commonwealth of Virginia has given to her sister States of this Union, has been that of an undertaking to compile and publish a complete collection of her *statutes at large;* that is, of all the acts of her legislative assemblies, from the first settlement of the colony to the present time. This work is, at this time, in the process of publication; and besides exhibiting the series of all the direct proceedings for the regulation of weights and measures, contains a mass of information, shedding light on every portion of our national history. The connection of weights and measures with the successive progress of this legislation, is more intimate and remarkable from the fact, that the original staple commodity of the colony, *tobacco,* was, for more than a century, not only merchandise, but *money.* It was the circulating medium of exchange; and

to a great degree so continued, until supplanted by the
modern and less valuable article of bank paper. To trace
the varieties of value, affixed to this article of tobacco, in
its character of a circulating medium, as rated by legislative enactments, in comparative estimation with other
articles of traffic, with the sterling currency of the mother
country, with foreign coins of gold, silver, and copper,
with the assessment of taxes, the levies of imposts, the
wages of labor, and the compensations for public service,
would be an inquiry into facts of high and interesting
curiosity, but too far transcending the immediate objects
of Congress, to be properly comprised in this report. It
must suffice to say, that the inspection laws relating to this
article would have been so numerous and so variant, that
the collection of them would alone fill several volumes. The
latest of these laws, and that which is now in force, is of
6th March, 1819, and provides, that the tobacco hogshead
shall not be more than 54 inches long of the stave, nor
more than 34 inches at the head within the crow, making
reasonable allowance for prizing, not exceeding two inches
above the gage in the prizing head; and that it shall contain 1250 pounds net of tobacco, with certain allowances
for shrinkage.

The assize of casks for other articles, as in most of the
other States in the Union, is regulated by different laws,
adapted to the different articles. The barrel for tar, pitch,
and turpentine, by an act of 26th December, 1792, must
contain $31\frac{1}{2}$ wine gallons, the precise nominal dimensions
of the old English wine barrel or half-hogshead, as pro-
scribed by acts of parliament time out of mind. But, by
the same act of 1792, barrels for beef and pork are to con-

tain 204 pounds net of meat, with an allowance of $2\frac{1}{2}$ per cent. for shrinkage; and are to be of capacity from 29 to 31 gallons. By another act of 28th December, 1795, barrels for fish are to be of not less than 30, nor more than 32 gallons. By an act of 8th January, 1814, the barrel of *salt* is to contain five bushels; agreeing thereby with the primitive Virginian corn barrel of 1631. But an act of 18th February, 1819, now requires that the barrels for bread, flour, or Indian meal, should be made of staves 27 inches long, and be of $17\frac{1}{2}$ inches diameter at the head, and contain 196 pounds of flour or meal.

The size of staves and heading is regulated by an act of 21st February, 1818, as follows:—

Staves—long butt, from 5 feet 6 inches to 5 feet 9 inches
from 5 to 6 inches broad,   [long,
from 2 to $2\frac{1}{2}$ inches thick.

Short butt and pipe, from 4 feet 6 to 4 feet 9 inches long,
from 3 to 4 inches broad,
from $\frac{5}{8}$ of an inch to $1\frac{1}{4}$ thick.

Hogshead—from 3 feet 6 to 3 feet 9 inches long,
from 3 to 4 inches wide,
from $\frac{3}{4}$ to $1\frac{1}{4}$ inch thick.

Barrel—from 2 feet 8 to 2 feet 10 inches long,
not less than 3 inches wide,
not less than $\frac{3}{4}$ of an inch thick in
any place.

Heading—of 28, 30, 32, in due proportion, and
not more than 34 inches long,
from 5 to 7 inches broad, dressed and
clean of sap, and from $\frac{3}{4}$ to $1\frac{1}{4}$
inch thick.

NORTH CAROLINA.

The only law of this State relating to weights and measures, a knowledge of which has been obtained, was enacted prior to the American Revolution, during the administration of Governor Gabriel Johnston, and is yet in force. It prohibits the use, in trade, by all the inhabitants or traders within the province, of any weights and measures other than are made or used *according to the standard in the English Exchequer,* and the statutes of England in that case provided. It charges the justices of the county courts to provide, at the charge of each county, sealed weights of half-hundred, quarter of hundred, seven pounds, four pounds, two pounds, one pound, and half-pound; measures of ell and yard, of brass or copper; measures of half-bushel, peck, and gallon, of dry measure, and a gallon, pottle, quart, and pint, of wine measure. It prescribes the appointment of standard-keepers in each county, to whom all weights and measures of the inhabitants are to be brought to be sealed, and who are to be sworn to the faithful discharge of their duties; and it subjects to suitable penalties the various offences of falsifying weights and measures, or of trading with such as have not been duly tried by the standard and sealed. It also repeals all former laws of the province upon the subject.

SOUTH CAROLINA.

By an act of 12th April, 1768, the public treasurer was required to procure, of brass or other proper metal, one weight of 50 pounds, one of 25 pounds, one of 14 pounds, two of 6 pounds, two of 4 pounds, two of 2 pounds, and two of 1 pound, avoirdupois weight, *according to the standard of London ;* and one bushel, one half-bushel, one peck, and one half-peck measures, *according to the standard of London.* The weights were to be stamped or marked in figures denominating their weight, and to be kept by the public treasurer; and by these weights and measures, declared to be the standards, all others in the province were to be regulated. By another act, of 17th March, 1785, subsequent to the Revolution, the justices of the county courts were authorized to regulate weights and measures within their respective jurisdictions, and to enforce the observance of their regulations by adequate penalties.

GEORGIA.

An act of the State legislature of 10th December, 1803, declares the standard of weights and measures established by the corporations of the cities of Savannah and Augusta to be the fixed standard of weights and measures within the State ; and that all persons buying and selling shall use that standard until the Congress of the United States shall have made provision on that subject. It directs the justices of the inferior courts, in the respective counties, to obtain standards conformable to those of the corporation of one of those cities; and prescribes regulations for keeping the standards, and for the trying, marking, and sealing, by them, the weights and measures of individuals, with penalties for using, in traffic, any others not corresponding with them.

An ordinance of the city council of Augusta directs that all weights for weighing any articles of produce, or merchandise, shall be of the *avoirdupois* standard weights; and all measures for liquor, whether of wine or ardent spirits, of the *wine* measure standards ; and all measures for grain, salt, or other articles usually sold by the bushel, of the dry or *Winchester* measure standard. And it prohibits the use of any other than brass or iron weights, thus regulated, or weights of any other description than those of 50, 25, 14, 7, 4, 2, 1, $\frac{1}{2}$, $\frac{1}{4}$, pound, 2 ounces, 1 ounce, and downwards.

KENTUCKY.

An act of the legislature, of 11th December, 1798, reciting in its preamble that Congress are empowered by the Federal Constitution to fix the standard of weights and measures, and that they had not passed any law for that purpose, recognizes, as thereby remaining in force within that commonwealth, the act of the General Assembly of Virginia, of the year 1734.

It therefore authorizes and directs the governor to procure one set of the weights and measures specified by the Virginian act of 1734, with measures of the length of one foot and one yard ; and declares that the bushel dry measure shall contain 2,150$\frac{2}{3}$ solid inches, and the gallon of wine measure 231 inches. It provides that these standards shall be kept by the secretary of state of the commonwealth ; that the governor shall cause to be made and transmitted to each county, scales and standards conformable to those of the State, which are to be kept by persons to be appointed by the county courts, and with which all the weights and measures, used in trade by individuals, are to be made to correspond.

TENNESSEE.

From a communication received from the governor of the State of Tennessee, it appears that there is in that State no standard of weights and measures fixed by the legislature.

OHIO.

The only act of the legislature of the State of Ohio, on this subject, is of 22d January, 1811. It directs the county commissioners of each county in the State to cause to be made one-half-bushel measure, to contain $1075\frac{2}{10}$ solid inches, which is to be kept in the county seat, and to be called the standard.

LOUISIANA.

Before the accession of Louisiana to the Union of these States, the weights and measures used in the province were those of France, of the old standard of Paris. An account of these, and of the present state of the weights and measures in the State of Louisiana, is submitted in the Appendix to this report.

By an act of the legislature of 21st December, 1814, the governor of the State was required to procure, at the expense of the State, weights and measures corresponding with those used by the revenue officers of the United States, together with scales and a seal, to be deposited in the custody of the secretary of the State, to serve as the general standard for the State.

Provision was also made by the same act for the appointment of an inspector at New Orleans, and for furnishing standards to the several parishes throughout the State.

By the last section of this act, a special dry measure is ordained by the name of a barrel, to contain three and a quarter bushels, according to the American standard, and to be divided in half and quarter barrel. The capacity of this measure, containing, according to the law, 6988.86 cubic inches, is referrible to none of the usual dry measures of the ancient Paris standard; but corresponds with tolerable exactness with the ancient Bordeaux half-hogshead, and with the assize of barrels prescribed by almost all the States of the Union, for packing beef, pork, and flour, for exportation.

INDIANA.

An act of the territorial legislature of 17th September, 1807, authorized the courts of common pleas of the respective counties in the territory, whenever they might think it necessary, to procure a set of measures and weights for the use of the county, namely, one measure of one foot, or twelve inches English measure, so called; one measure of three feet, or thirty-six inches English measure; one half-bushel for dry measure, to contain $1075\frac{1}{2}$ solid inches; one gallon measure, to contain 231 solid inches; the measures to be of wood, or any metal, as the court may think proper; also, one set of avoirdupois weights, to be sealed with the name or initial letters of the county. These weights and measures were to be kept by the clerks of the county courts for the purpose of trying and sealing those used in their counties. After due notice given by the courts that these standards had been procured, all persons were prohibited from buying or selling by weights or measures not corresponding with them: and the clerk was to try and seal all weights or measures brought to him therefor corresponding with the standard. This act was to continue in force till Congress should otherwise provide.

The provisions of this act are, in substance, and nearly to the letter, repeated in an act of the State legislature, of 21st January, 1818.

There is also an act of 24th December, 1816, regulating the inspection of tobacco, and one of 2d January, 1819, regulating the inspection of flour, beef, and pork. The assize of hogsheads and casks prescribed in them is the same as that of the Virginia laws.

MISSISSIPPI.

An act of the territorial legislature of 4th February, 1807, directed the treasurer to procure a set of the large avoirdupois weights according to the standard of the United States, if one were established; but if there were none such, according to the standard of London, with proper scales for weights, together with measures of foot and yard, dry measures of capacity, and liquid *wine* measures. He was required to furnish each county in the territory with a set of weights, scales, and measures, conformable to the above standards, to be kept by a person appointed by the county courts, under oath, and accessible to all persons desirous of having their weights and measures tried and scaled. Penalties were also annexed to the use of weights and measures not corresponding with these standards.

A subsequent act of 23d December, 1815, further required of the treasurer to procure six sets of the weights and measures as above described, and to distribute them at suitable places in the several counties of the territory; and additional penalties were prescribed for the use of weights and measures not corresponding with the standard.

An act of the legislature of the State of Mississippi, of 6th February, 1818, "to provide for inspections, and for other purposes," contains many other regulations for the keeping of the standard weights and measures, and for securing conformity to them. It makes no alteration of the standard, but confirms, "until Congress shall fix a standard for the United States," that which had already been established. It also requires that barrels of flour should contain 196 pounds net, and barrels of pork and beef 200 pounds net of meat.

ILLINOIS.

The territorial act of 17th September, 1807, passed while the State of Illinois formed a part of the Indiana territory

But by an act of the legislature of this State, "regulating weights and measures," of 22d March, 1819, the county commissioners of each county in the State were required to procure, at the expense of the county, one foot and one yard English measure; a gallon liquid or wine measure, to contain 231 cubic inches; corresponding quart, pint, and gill measures, of some proper and durable metal; a half-bushel dry measure, to contain eighteen quarts, one pint, and one gill, wine measure, or 1075.2 cubic inches, and a gallon dry measure, to contain one-fourth part of the half-bushel, these two measures to be of copper or brass; also, a set of weights of one pound, one half-pound, one eighth-pound, and one sixteenth-pound, made of brass or iron, the integer of which to be denominated one pound avoirdupois, and to equal in weight 7,020 grains troy, or gold weight. These weights and measures are to be kept by the clerk of the county commissioners for trying and sealing the measures and weights in common use.

All persons are authorized to have their weights and measures tried by the standards, and sealed, and are forbidden, upon suitable penalties, to buy or sell by others not corresponding with them.

The most remarkable peculiarity of this act is its departure from the English standard weights by fixing the avoirdupois pound at 7,020, instead of 7,000 grains troy.

ALABAMA.

This State having formed a part of the Mississippi territory, previously to the admission of the State of Mississippi into the Union in 1817, the acts of that territory of 4th February, 1807, and 23d December, 1815, embraced this section of territory. No act of the State legislature of Alabama, on this subject, is known to have been passed.

MISSOURI.

The territorial legislature, by an act of 28th July, 1813, directed the several courts of common pleas within the territory, to provide for, and at the expense of the respective counties, one foot and one yard English measures; one half-bushel, to contain $1075\frac{1}{2}$ solid inches, for dry measure; one gallon, to contain 231 solid inches, and smaller liquid measures in proportion, to be of wood, or any metal the court should think proper; also, one set of avoirdupois weights, and one seal, with the initial of the county inscribed thereon, all to be kept by the clerks of the courts of common pleas, or circuit courts, for the purposes of trying and sealing the measures and weights used in their counties.

The use, or keeping to buy or sell, of weights or measures not corresponding with these standards, after due notice, was prohibited under penalties by the same acts; but with a proviso, that all contracts or obligations, made previous to the taking effect of the act, should be settled, paid, and executed, agreeably to the weights and measures in common use when the contracts or obligations were made or entered into.

DISTRICT OF COLUMBIA.

By the act of Congress of 27th February, 1801, concerning the District of Columbia, the laws of the State of Virginia, as they then existed, were continued in force in the part of the District which had been ceded by that State, and the laws of Maryland in the part of the District ceded by Maryland.

The act to incorporate the inhabitants of the city of Washington, of 3d May, 1802, authorizes the corporation to provide for the safe keeping of the standard of weights and measures fixed by Congress, and for the regulation of all weights and measures used in the city.

The supplementary act, of 24th February, 1804, gives the city council power to establish and regulate the inspection of flour, tobacco, and salted provisions; and the gauging of casks and liquors.

And by the act of 4th May, 1812, further to amend the charter of the city of Washington, further power is given to the corporation to regulate the measurement of, and the weight by which, all articles brought into the city for sale shall be disposed of.

The weights and measures of the city have, accordingly, been regulated by various acts of the corporation, conformably to the standard used in the State of Maryland. The inspection laws, the assize of tobacco hogsheads and flour casks, the dimensions of bricks and of cord-wood, are all formed upon the same model. The weight of bread is adapted once a month to the price of flour: but by a special ordinance, all coal for sale within the city is sold by a measure containing five struck-standard half-bushels, stamped and marked by the sealer of weights and measures, and the stricken measure of which is considered as two bushels.

As preliminary remarks, in reference to that part of the resolutions of both Houses which requires the opinion of the Secretary of State with regard to the measures which it may be proper for Congress to adopt in relation to weights and measures, it may be proper to state the extent of what can be done by Congress. Their authority to act is comprised in one line of the Constitution, being the fifth paragraph of the eighth section and first article; in the following words: *"to fix the standard of weights and measures."*

It may admit of a doubt whether under this grant of power is included an authority so totally to subvert the whole system of weights and measures as it existed at the time of the adoption of the Constitution, as would be necessary for the introduction of a system similar to that of the French nation. To *fix* the standard, appears to be an operation entirely distinct from changing the denominations and proportions already existing, and established by the laws, or immemorial usage. And this doubt acquires a further claim to consideration, if it be true, as the experience of other nations seems to warrant us in the conclusion, that there is no object of regulation by human power, in which the prescriptions of a government are so difficult to be carried into execution. Throughout Europe, in the most absolute as well as in the freest governments, every historical research presents a fruitless struggle on the part of authority to introduce order and uniformity: and an unconquerable adherence of custom to the diversities of usage among the people. There is perhaps less of this diversity in the United States than in any country in Europe. At the adoption of the Constitution all the weights and measures in common use throughout the United States were derived, either by the statutes of the States, or by an invariable usage, which had supplied the place of law, from the standards in the English Exchequer. Hence, the English foot, divided into twelve inches, was the unit of all measures of matter in length, breadth, or thickness. Its various multiples of the yard, ell, perch, pole, furlong, acre, and mile, were all recognized by the laws, and in the familiar use of the people. The avoirdupois and troy weights, with the difference of modification of the latter as used for weighing the precious metals or apothecary's drugs in retail, the wine gallon of 231, and the beer gallon of 282 solid inches. were equally well known, and in general use. and the Winchester bushel, of

2,150.42 solid inches, formed the general standard of all the dry measures of capacity.

In many of the States the standards established by statute had been procured from the Court of Exchequer; and the only variety discernible in the legislation of the States on this subject, arises from a difference existing in the several standards of the same measures at the Exchequer, and at Guildhall in London.

In the exercise of the authority of Congress, with a view to the general principle of uniformity, there are four different courses of proceeding which appear to be practicable.

1. To adopt, in all its essential parts, the new French system of weights and measures, founded upon the uniformity of identity.

2. To restore and perfect the old English system of weights, measures, moneys, and silver coins, founded upon the uniformity of proportion.

3. To devise and establish a system, in which the uniformities of identity and of proportion shall be combined together, by adaptations of parts of each system to the principles of the other.

4. To adhere, without any innovation whatever, to our existing weights and measures, merely fixing the standard.

1. In the review which has been taken, and the comparison which has been submitted to Congress, between the old English, and the new French, or as they may with more propriety be called, the ancient and the modern systems of metrology, it has been the endeavor of this report to show, that, while each of these systems embraces principles of the highest importance, neither of them includes

all the elements resulting from the nature of the relations
between man and things as created beings, and between
man and man in society, mingling in the purposes to
which weights and measures are applicable. The opinion
has been expressed, that the uniformity of proportion in
the ancient system, uniting weight and measure by the
relative gravity, extension, and numbers, incident to dry
and liquid substances, possessed advantages, of which the
uniformity of identity in the modern system was entirely
deprived; that the property of the ancient system, by
which the money weight and the silver coin were the same,
the most useful of all uniformities of which weights, measures, money, and coins are susceptible, was very imperfectly
adapted to the modern system of France; that the French
system, admirable as it is, looked, in its composition, to
weights and measures, more as exclusively matters of account, than as tests of quantity; that, in its eagerness for
extreme accuracy in the relations between things, it lost
sight a little of the relations of weight and measures with
the physical organization, the wants, comforts, and occupations of man; that, in its exclusive partialities for decimal
arithmetic, it forgot the inflexible independence and the
innumerable varieties of the forms of nature, and that she
would not submit to be trammelled for the convenience of
the counting-house. The experience of the French nation
under the new system has already proved, that neither the
immutable standard from the circumference of the globe,
nor the isochronous vibration of the pendulum, nor the
gravity of distilled water at its maximum of density, nor
the decimation of weights, measures, moneys, and coins,
nor the unity of weight and measure of capacity, nor yet

all these together, are the only ingredients of practical uniformity for a system of weights and measures. It has proved, that gravity and extension will not walk together with the same staff; that neither the square, nor the cube, nor the circle, nor the sphere, nor the revolutions of the earth, nor the harmonies of the heavens, will, to gratify the pleasure, or to indulge the indolence of man, be restricted to computation by decimal numbers alone.

The substitution of an entire new system of weights and measures, instead of one long established and in general use, is one of the most arduous exercises of legislative authority. There is indeed no difficulty in enacting and promulgating the law; but the difficulties of carrying it into execution are always great, and have often proved insuperable. Weights and measures may be ranked among the necessaries of life, to every individual of human society. They enter into the economical arrangements and daily concerns of every family. They are necessary to every occupation of human industry; to the distribution and security of every species of property; to every transaction of trade and commerce; to the labors of the husbandman; to the ingenuity of the artificer; to the studies of the philosopher; to the researches of the antiquarian; to the navigation of the mariner, and the marches of the soldier; to all the exchanges of peace, and all the operations of war. The knowledge of them, as in established use, is among the first elements of education, and is often learnt by those who learn nothing else, not even to read and write. This knowledge is rivetted in the memory by the habitual application of it to the employments of men throughout life. Every individual, or at least every family, has the weights and measures used in the vicinity, and recognized by the custom of the place. To change all this at once, is to affect the well-being of every man, woman, and child, in the community. It enters every house, it cripples every hand. No legislator can attempt it with any prospect of success, or any regard to justice, but upon two indispensable conditions: one, that he shall furnish every individual citizen easy access to the new standards which take the place of the old ones; and the other, that he shall enable him to *know* the exact proportion between the old and the new. A multiplication of standard copies to a great extent is indispensable; and the distribution of them throughout the country, so that they may be within the means of acquisition to every citizen, is among the duties of the government undertaking so great a change. Tables of equalization must be circulated in such manner as to find their way into every house; and a revolution must be effected in the use of books for elementary education, and in all the schools where the first principles of arithmetic may be taught. All this has been done in France; and all this might be done perhaps with more ease in the United States. But, were the authority of Congress unquestionable to set aside the whole existing system of metrology, and introduce a new one, it is believed that the French system has not yet attained that perfection which would justify so extraordinary an effort of legislative power at this time.

The doubts entertained whether an authority, so extensive as this operation would require, has been delegated to Congress, are strengthened by the consideration of the character of the executive power, corresponding with the legislative authority. The means of execution for exacting and obtaining the conformity of individuals to the ordinances of the law, in the case of weights and measures, belong to that class of powers which, in our complicated political organization, are reserved to the separate States. The jurisdictions to which resort must be had for transgressions of this description of laws, are those of municipal police. In England they were originally of the resort of views of frank-pledge in every separate manor, and have since been transferred to the clerks of the market and to the justices of the peace. The sealers of weights and measures, officers who have the custody of the standards, and the authority to compare with them, from time to time, the weights and measures used by individuals, and to prosecute for all offences by variations from the standards, and the courts before whom all such offences are triable, are institutions not only existing in almost every State in the Union, but essentially belonging to that portion of public authority suited to the State administration rather than to that of the Union. It is a general principle of our constitutions, that, with every delegation of legislative authority, a co-extensive power of execution has been granted. Affairs of municipal and domestic concern have, for obvious reasons, been reserved to the State authorities; and of this character are most of the regulations and penal sanctions for securing conformity to the standards of weights and measures. In *fixing the standard*, it is believed that Congress must rely almost entirely, if not altogether, upon State executive authorities for carrying their law into execution. And, although this reliance may be safely indulged in relation to a law which should merely fix the uniformity of existing standards, its efficacy would be very questionable in the case of a law of great and universal innovation upon the habits and usages of the people. Of such a law the transgressions could not fail to be numerous: any doubt of the authority of the legislator would stimulate to systematic resistance against it: and the power of enforcing its execution being in other hands, naturally disposed to sympathize with the offender, the whole system would fall into ruin, and afford a new demonstration of the impotence of human legislation against the laws of nature, in the habits of man.

2. The restoration of the old English, which was also the Greek and Roman, system of weights, measures, and silver coins, founded upon the uniformity of proportion, would require an exercise of authority no less transcendent than the introduction of the French system. Its advantages were, the identity of the money weight and silver coin, the wine gallon at once a multiple of the money weight, and an aliquot part of the cubic foot; and its proportions of the money and commercial pounds, and of the wine and corn gallons, to the relative specific gravity of wine and wheat. But, as all these combinations were founded upon the assumption that the relative gravity of wheat to wine was as 4 to 5, and that the gravity of wine and of spring-water was the same; and as it allowed of the making of the wine gallon by the two processes, by the weight of wheat multiplied, and by the weight of the cubic foot of water divided, the result of the two processes was not exactly the same. The Irish gallon, of 217.6 inches, was made by one process; and the Rumford gallon, of 266.25, was its corresponding corn and ale measure. A

wine gallon of 219.5 cubic inches was made by assuming 252 gallons as the measure of the ton, or 32 cubic feet; and its corresponding corn measure was the Winchester bushel, with an ale gallon of 268. The Winchester bushel is the only existing relict of the old English system, which has outlived all the changes of the laws, and all the revolutions of ages. Should that be retained, and its contents fixed at 2,148.5, to restore and perfect the whole system by an exact combination of the two modes of forming the water gallon, without regard to the weight of wine, would require a liquid gallon of 219.5 inches, a dry gallon of 268.5, a money pound of 5,714.28, and a commercial pound of 6,944.44 grains troy. This money pound should then be made the weight of the unit of silver coins, of a settled standard purity, and might be decimally divided, like our present silver coins, and decimally or duodecimally divided as a weight. Or, the ton might be declared to contain 256 gallons, of 216 cubic inches; in which case the money pound would be 5,625, and the commercial pound of 6,836 grains troy; the corn and ale gallon of 262.5, and the bushel of 2,100 cubic inches. If the old easterling 12 and 15 ounce pounds should be restored, and the gallon, according to its primitive composition, be made to contain ten 12-ounce pounds of wine, it would then be, considering the gravity of wine as of 250 grains troy to a cubic inch, of the same capacity of 216 cubic inches. It would also contain eight 15-ounce pounds, of 6,750 grains troy; but the proportion between the two pounds would not be exactly that between the gravity of wheat and wine. The wine gallon, filled with eight 12-ounce pounds of wheat, would contain, in wine, eight pounds, not of 6,750, but of 6,608 grains, and, if divided into fifteen ounces, the ounce would not be the easterling, but the avoirdupois ounce.

3. The proportions between the existing troy and avoirdupois weights, and between the wine gallon of 231, and the beer gallon of 282 cubic inches, are more exactly those between the specific gravity of wheat and of spring-water, than were the easterling pounds of 12 and 15 ounces, or those of the primitive gallon of 216 inches with the ale gallon deduced from the Winchester bushel. They are exact, to the utmost degree of precision; but these proportions are without use. Neither does the wine gallon contain an exact number of pounds of wine, nor is the beer gallon an aliquot part of the bushel. These were proportions, in their origin, of great usefulness, but imperfectly settled. The whimsical operation of time and human laws upon them has been to make the proportions perfect, but to render them useless. There are, nevertheless, very useful proportions in our existing weights and measures, one of which is between the ton measure of water and the pound avoirdupois. As 1000 ounces avoirdupois weigh exactly one cubic foot of water, it follows that the ton of 2,000 pounds weight is the ton of 32 cubic feet measure. The other is between the pound avoirdupois and the pound troy; the former consisting of precisely 7,000 grains troy. The pound avoirdupois is therefore the connecting link between weight and linear measure. It is at once a test and standard of the cubic foot, of the ton measure, and of the troy weight; while the foot, the ton, and the troy weight are each, by this connecting link, tests and standards of each other, and of the avoirdupois pound. But the thirty-two cubic feet, which are at once the ton weight of

two thousand pounds, and the ton measure of water, are not sufficient, as measure, to contain the same weight of wheat. The bushel is the measure containing the same weight of wheat which the cubic foot contains of water. Thirty-two bushels, therefore, contain the ton weight, of two thousand pounds avoirdupois; but they would make a ton measure within a small fraction of 39 cubic feet.

The avoirdupois pound of 16 ounces, and of 7,000 grains troy, is used, however, only for quantities of less than a quarter of a hundred pounds. It then receives an accession of 12 per cent. on its quantity; the quarter of a hundred contains 28 pounds, the hundred 112, and the ton of 2,000 actually contains 2,240. If the hundred and twelve pounds should be considered as a net hundred, each pound would be of 7,840 grains troy weight, and would bring it within one-quarter of an ounce troy to the weight of the French half-kilogramme, or usual pound. If the wine gallon were, as under the statute of 1496 it should have been, and as the Guildhall gallon before the statute of 5 Anne actually was, of 224 inches, it would have had two further useful coincidences: it would have contained just eight pounds avoirdupois of wine, eight pounds troy weight of wheat, and a number of cubic inches in decimal subdivision to the number of pounds avoirdupois in the ton of 2,240, or twenty hundred of 112 pounds.

There are two changes, therefore, in our existing weights and measures, which would restore and perfect the system of ancient metrology; one, to make the troy weight the unit of our silver coins, in which case it might be decimally divided as coin, retaining its divisions into ounces, penny-weights, and grains, as a weight; and the other, to restore the wine gallon of 224 inches, with its corresponding ale gallon of 272, and bushel of 2,176 inches.

But it has been already remarked, that in the ancient system, founded on the uniformity of proportion between the relative extension and gravity of wheat and wine, there were, in the double sets of weights and measures of capacity, two advantages; one, of a general nature, resulting from it as proportional, without reference to the articles selected for settling the proportions; and the other special, arising from the selection of wheat and wine as the articles. The first belongs to every proportional system of which the proportion between the standards is accurately ascertained, and consists in this, that each weight and each measure is a test and standard for all the others. The second depends on the selection of the articles, and is limited to the conveniences and facilities of trade, commerce, and navigation, as incidental to them. Reasons have been suggested, why the two articles of wheat and wine should have been selected in the primitive system, as being, from the nature and physical constitution of man, the first, and, for many ages, the greatest and most important articles of traffic. The necessity for establishing a proportion between the relative weight and measure of those articles, was also dictated by the practice of transporting them both by sea in ships.

The space in cubic feet which would be filled by a determinate weight of each of them was an object of essential importance to be known, not as a philosophical theory, but for every mechanical operation of the commerce. The size of the cask must be adapted to the capacity and the burden of the ship; and when the ton weight of wine had been adapted to the ton measure of water, it became of the utmost use to make the measure of corn so correspond with the cask of wine, as to contain the same determinate quantities by weight. But in modern times, and especially to these United States, neither wheat nor wine is an article of primary importance in domestic trade, or in foreign commerce. Whatever may be the capacities of our country for producing wine, they have hitherto scarcely been discovered. Tea and coffee have taken the place of wine as comforts, or next to necessaries of life; and have degraded that article into the class of luxuries. We import little, and export none of it. We receive it in the casks of the several countries from which it comes; and although the laws of some of our States, as well as those of England, still exhibit the absurdity of requiring that the hogshead should contain 63 wine gallons of 231 cubic inches, because it once contained 63 gallons of 219½ inches, yet no one complains that the real hogshead is just what it was 600 years ago, without either swelling to the dimensions of Queen Anne's cubic inches, or contracting the gravity of its contents to the troy weight of Henry the Seventh. We raise vast quantities of wheat, but export it almost exclusively in its manufactured state of flour. The weight of wine is, between the buyer and seller, never a subject of inquiry. We have universally the Winchester bushel, defined by the 13 William III., of 2,150.42 cubic inches, with the single exception of the State of Connecticut, whose standard bushel is *very near* 2,198 inches. And the laws of many of the States require, that the bushel should contain 60 pounds avoirdupois of wheat. Should a standard bushel now be made in the manner described in the statute of 1266, it would be a measure of 2,148.5 cubic inches, and would contain 60¼ avoirdupois pounds of wheat. The relative proportion between the extension and specific gravity of wheat and wine is to us, therefore, of no importance or use in our system of weights and measures. When the wine gallon contained a determinate weight of the liquor, and was at the same time a sixty-third part of eight cubic feet, there were motives of convenience and utility in using another measure for ale and beer, which, being brewed from grains, had natural proportions to the measures used for them. It was natural, therefore, to employ the eighth part of the measure of the bushel as the beer gallon, though at the same time a vessel of smaller size was used for the measurement of wine. But since the weight of wine, and the proportions of its measuring vessel to the cubic foot, have ceased to be of any account, there is no purpose of utility answered by the employment of two different measures for different fluids; while there is great tendency to error and fraud in the use of two such measures, of the same materials and bearing the same name.

4. Our system of weights and measures is, therefore, susceptible of great improvements, by restoring some of the principles which belonged to the system from which it was originally derived. It is perhaps still more improvable, by the adoption of some of the principles contained in the new French metrology. There is no doubt that the decimal divisions might be introduced to great advantage both into linear measure by the adoption of the metre, and into weights, by identifying the money weight with the silver coin. It is believed that a system, embracing the essential

advantages of all the three, might, without much difficulty, be combined; and that it would be better adapted than either of them to the use of all human kind, and thus secure, in its utmost possible extent, the uniformity with reference to persons.

Weights and measures, and the final establishment of a system for them, with a view to the almost practicable extent of uniformity, are at this moment under the deliberative consideration of four populous and commercial nations —Great Britain, France, Spain, and the United States. The interest is common to them all: the object of *uniformity* is the same to all. Could they agree upon one result, the advantages of that agreement would be great to each of them separately, and still greater in all their intercourse with one another. But this agreement can be obtained only by consultation and concert. It is, therefore, respectfully proposed, as the foundation of proceedings necessary for securing ultimately to the United States a system of weights and measures which shall be common to all civilized nations, that the President of the United States be requested to communicate, through the ministers of the United States in France, Spain, and Great Britain, with the governments of those nations, upon the subject of weights and measures, with reference to the principle of uniformity as applicable to them. It is not contemplated by this proposal that the communication should lead to any conventional stipulations or treaties, but it is hoped that the comparison of ideas, and the mutual reciprocation of observation and reflection may terminate in concurrent acts, by which, if even universal uniformity should be found impracticable, that which would be obtained by each nation would at least approximate nearer to perfection.

In the mean time, should Congress deem it expedient to take immediate steps for accomplishing a more perfect uniformity of weights and measures within the United States, it is proposed that they should assume as their principle, that no innovation upon the existing weights and measures should be attempted.

To fix the standard of weights and measures of the United States as they now exist, it appears that the act of Congress should embrace the following objects:

1. To *declare* what are the weights and measures to which the laws of the United States refer as the legal weights and measures of the Union.

2. To procure positive standards of brass, copper, or such other materials as may be deemed advisable, of the yard, bushel, wine and beer gallons, troy and avoirdupois weights, to be deposited in such public office at the seat of government as may be thought most suitable.

3. To furnish the executive authorities of every State and territory with exact duplicates of the national standards deposited at the seat of government.

4. To require, under suitable penal sanctions, that the weights and measures used at all the custom-houses, and land surveys, and post-offices, and generally by all officers under the authority of the United States in the execution of their laws, should be conformable to the national standards.

5. To declare it penal to make or to use, with intent to defraud, any other weights and measures than such as shall be conformable to the standards.

1. The existing weights and measures of all the States of this Union are derived from the Exchequer, or from the laws of Great Britain. The one common standard from which they are all deduced is the English foot, divided into twelve inches, and three of which constitute the yard. The positive standard yard is a brass rod of the year 1601, in the British Exchequer. The unit of measure is the foot of twelve equal inches. The inch, by the English laws, is divided into three equal parts, called barley-corns, but this division is not used in practice. The practical divisions of the inch are, at option, binary, or decimal: that is, of halves, quarters, and eighths, or of tenths, hundredths, and thousandths. Thirty-two cubic feet of spring-water, at the temperature of 56 degrees of the thermometer of Fahrenheit, constitute the ton weight of two thousand pounds avoirdupois. The pound avoirdupois consists of sixteen ounces; the ounce, of sixteen drachms. The pound avoirdupois is equal in weight to seven thousand grains troy, or to fourteen ounces, eleven pennyweights, sixteen grains, troy. The troy pound consists of twelve ounces, each ounce of twenty pennyweights, each pennyweight of twenty-four grains. It is otherwise divided for the use of apothecaries; but the grain and the pound are the same. The troy pound is equal in weight to 13 ounces and $2\frac{3}{5}$ drams avoirdupois.

The bushel is a cylindrical vessel $18\frac{1}{2}$ inches in diameter, and eight inches deep, or any vessel of 2,150.42 cubic inches. It is divided into four pecks, each peck into four pottles, each pottle into two quarts, each quart into two pints.

The ale and beer gallon is a vessel of 282 cubic inches.

It is divided into four quarts, each quart into two pints, each pint into four gills.

The wine gallon is a vessel of 231 cubic inches, divided, like the beer gallon, into wine quarts, pints, and gills.

Any cubic vessel of 12.9 inches in length, breadth, and thickness, is of equal contents with the Winchester bushel. Any cubic vessel of 6.55767, is of equal contents with the ale gallon. Any cubic vessel of 6.13579 is a wine gallon.

For the purposes of the law, it will be sufficient to declare that the English foot, being one-third part of the standard yard of 1601 in the Exchequer of Great Britain, is the standard unit of the measures and weights of the United States; that an inch is the twelfth part of this foot; that thirty-two cubic feet of spring-water, at the temperature of 56 degrees of Fahrenheit's thermometer, constitute the ton weight of 2,000 pounds avoirdupois; that the gross hundred of avoirdupois weight consists of 112 pounds, the half-hundred of 56, and the quarter-hundred of 28, the eighth of a hundred of 14, and the sixteenth of a hundred of 7 pounds; that the troy pound consists of 5,760 grains, 7,000 of which grains are of equal weight with the avoirdupois pound; that the bushel is a vessel of capacity of 2,150.42 cubic inches, the wine gallon a measure of 231, and the ale gallon a measure of 282 cubic inches.

The various modes of division of these measures and weights, the ell measure, and the application of the foot to itinerary, superficial, and solid measure, producing the perch, rood, furlong, mile, acre, and cord of wood, may be left to the established usage, or specifically declared, as may be judged most expedient. The essential parts of the

whole system are, the foot measure, spring-water, the avoirdupois pound, and the troy grain.

2. For the purpose of uniformity, it would be desirable
to obtain a copy, as exact as the most accomplished art
could make it, of the standard yard of 1601, in the Exchequer of Great Britain, made of the same material,
brass, but divided with all practicable accuracy into three
feet, and thirty-six inches, and each inch further divided
into tenth and hundredth parts. This rod, with the words,
"standard yard measure of the United States—three feet
—thirty-six inches:" and the date of the year engraved on
one of its sides, should be enclosed in a wooden case, and
deposited for safe-keeping in one of the offices at the Cap-
itol. From the foot measure of this yard, the standard
bushel, and two gallons, should be made. The avoirdupois pound, and the troy weight of 256 ounces, should be
made exactly conformable to the standards in the Exchequer. The weights of 56, 28, 14, and 7 pounds avoirdupois, should be made exact multiples of the pound
weight. But no subdivisions of the bushel or gallons,
or of the avoirdupois pound, should be placed among the
standards. An enactment, that no subdivisions of the
standards, other than in the due proportion to them,
should be legal, would avoid the inconvenience and the
varieties which multiplied material standards always produce. All the standards should, like the yard, have their
names, as standards of the United States, the date of the
year, and a designation of quantity engraved upon them.
On the bushel, for instance,—" 2,150$\frac{42}{100}$ cubic inches;" on
the wine and ale gallons, respectively, 231 and 282 inches;
on the avoirdupois pound " 7,000 grains troy weight, avoir-

dupois pound," on the troy weights "256 ounces—12 ounces, and 5,760 grains to the pound troy weights." These standards, all enclosed in suitable cases, to preserve them from injury, and, as effectually as possible, from decay, should be deposited in the custody of a sworn and responsible officer, with the standard yard.

3. These national standards being thus made and deposited, exact copies of them should be made of the same materials, substituting for the words "standard of the United States," engraved upon the originals, the words "United States" standard, State "of—:" and these copies should be transmitted to the executives of every State in the Union. The standard for the territories might leave the name of the State to be engraved when the territory should pass to that condition: and the standards for the District of Columbia might properly be committed to the charge of the clerk of the Supreme Court of the United States.

4. It should be made the duty of the collectors, surveyors, and naval officers of the customs, the registers of the land offices, and receivers of public moneys, of the postmaster-general, and all postmasters, the quartermasters, and commanding officers at military posts of the army, the commanding officer and purser of every vessel of the navy, the commanding officer at the military academy, of all Indian agents, and of the marshals of the several judicial districts of the United States, to ascertain, and to certify in writing, upon oath, to the heads of their respective departments, that the weights and measures used by them, in the discharge of their official duties, are conformable to the standards of the United States. And to secure

the future observance of this uniformity, every such officer, civil or military, to be appointed hereafter, should, together with the oath to support the Constitution of the United States, have administered to him an oath that he will, in the discharge of his official duties requiring the employment of weights and measures, scales and beams, use such as are conformable to the legal standards of the United States, and not knowingly any others. To the penalties of removal from, and disqualification for office, might be added a right of action for damages, given to any person injured by the wilful neglect or refusal of any such officer to observe the requisitions of the law.

5. The offence of fraudulently or wilfully making or selling any weight, measure, scales, or beam, to be used as conformable to the United States' standards, and not conformable, might be made punishable by fine and imprisonment, upon presentment and conviction before the circuit courts of the United States.

The existing laws of all the States should be declared, so far as they are conformable to the act of Congress *fixing* the standard, to remain unrepealed and in full force. All sealers of weights and measures, and all persons appointed under the authority of the several States for the custody of standards, should be required to ascertain them to be conformable to the standards of the United States. It is scarcely possible that any law of the United States to establish uniformity of weights and measures throughout the Union, should be made effectual, without the cordial aid and co-operation of the State legislative and executive authorities. This is one of the most powerful reasons which have led to the conclusion, that, in fixing the standard, all present innovation should be avoided. The standards of all the States are now, or by their laws should be, the same as those herein proposed, excepting only the Connecticut bushel, the change in which will be inconsiderable. Several of the States have systems well organized, and in full operation for the uniformity of their weights and measures. The standards of many of them are incorrect; some from careless usage and decay; others from having been copies of copies made without much attention to accuracy; and, others from having transferred to this country all the varieties of the original standards in the Exchequer. The object of the act, the substance of which is now proposed to Congress, would be, to make the uniformity already existing by the laws and usages of every part of the Union more effectual and perfect in point of fact. The table of a return from the several customhouses of the United States will show the extent of the existing varieties; and while they add new demonstration of the justness of the sentiment universally prevailing, that the authority delegated to Congress by the Constitution, of fixing the standard, should be exercised without delay, they also show that the best exercise of that authority will be by making it essentially auxiliary to the efficacy of the existing State laws.

In the consultation which it is proposed that the President of the United States should be requested to authorize and conduct with foreign governments, with a view to future, more extensive, and perfect uniformity, there is one object, which, it is presumed, may be accomplished with little difficulty or expense, and by means of which the standard from nature of the new French system, the metre, may be engrafted upon our system without discomposing any of its existing proportions.

In all the proceedings, whether of learned and philosophical institutions, or of legislative bodies, relating to weights and measures within the last century, an immutable and invariable standard from nature of linear measure has been considered as the great desideratum for the basis of any system of metrology. It is one of the greatest merits of the French system to have furnished such a standard for the benefit of all mankind, in the metre, the ten millionth part of the quarter of the meridian. Of the labors, and researches, and liberal expense, and art, and genius, which have been lavished by France upon this operation, and of the success with which it has been accomplished, the notice which it amply merited has already been taken in this report. Since this great and admirable undertaking has been achieved, a disposition to detract from its merit and usefulness has been occasionally manifested. Some philosophical speculators have started doubts whether the metre is really the forty millionth part of the circumference of the earth; and indeed whether such a measure can, with perfect accuracy, be ascertained by human art. Other standards from nature have been suggested as preferable to the arc of the meridian: individual passions and anti-social prejudices have insinuated themselves into the inquiry: and the question between the metre and the pendulum has almost festered into a test of party controversy, and an engine of national jealousy. In the establishment of the French system, the pendulum, as well as the meridian, has been measured; but the *standard* was, after long deliberation, after a cool and impartial estimate of the comparative advantages and inconveniences of both, definitively assigned to the arc of the meridian, in departure from an original prepossession in favor of the pendulum. Two reasons are deemed decisive for concurring in the principle of this determination; one, that the earth being the greatest object of actual measurement within the physical powers of man, an aliquot part of its circumference is the only measure, which, applicable to that object, is also equally applicable to every other purpose of weight or mensuration; and the other, that this standard once settled is invariable, while the pendulum, being of different lengths in different latitudes, is essentially defective in one of the most important principles of uniformity, that of *place* or capacity of application to every part of the earth.

It is proposed, therefore, to discard all consideration of the pendulum: as the theory of its vibrations, however interesting in itself, is believed to be, since the definitive determination of the metre, useless, with reference to any system of weights and measures. Nor is it of more importance to know whether the metre really be, within the ten thousandth part of an inch an exact aliquot part of the circumference of the earth. An error to that, or even to a greater extent, admitted to be possible, leaves for all practical purposes of human life, even including the operations of geography and astronomy, the metre as perfect a standard for weights and measures as any other that ever was devised, and a much more perfect one than the pendulum.

It is therefore submitted to the consideration of Congress, that, in the act for fixing the standard of weights and measures for the United States, together with a definition of the foot, its exact proportion to the standard metre of France should be declared: to effect which purpose with the utmost attainable accuracy, it would be necessary to compare together the identical measure, to be used hereafter as the standard linear measure of the Union, with the standard metre in platina, deposited in the national archives of France. It is not doubted that the French government would readily give their assent to this operation, and would agree that it should be performed in such manner as to settle, definitively, for the future use of both countries, the exact proportion to the ten thousandth part of an inch, between the foot measure of the United States and the metre. From the perfection which the instruments used for comparing together measures of length have attained, accuracy to that extent may be effected. But the necessity of such an operation for the definitive settlement of this proposition is apparent, from the fact, that the comparisons hitherto made in France, and in England, and in the United States, though all made with all possible care, have terminated in results so different, that it would scarcely be safe to assume either of them as the proportion to be declared by a legislative act.

In the attempt to determine distances of space less than the 200th part of an inch, the experiment is met by obstacles, in the temperature and pressure of the atmosphere, and in the different degrees of their influence upon the matter to be measured. Heat and cold, moist and dry, high and low, affect the metals of which measures are composed with various degrees of dilatation and contraction. Brass, the metal of which the English standards are formed, being a compound metal, is variously dilatable: and, although tables have been formed of the degrees in which the simple metals are expanded by heat, according to the scale of the thermometer, yet, as those tables, made by different men, do not agree, no perfect reliance can be had upon them. As yet no experiments of admeasurement, made by different persons, at different times, but of the same standards, have exhibited results, approximating within one two-hundredth part of an inch; of a contrary result, the examples are numerous, and so remarkable that they deserve to be noticed more particularly.

In the year 1797, Sir George Shuckburg Evelyn measured, with Troughton's microscopic beam compass and scale, all the standards at the Exchequer; the scale made by Sisson for Graham in 1742, the parliamentary standards of 1758 by Bird, the scale used by General Roy for the measurement of the base, and several others. The result of his experiment was published in the transactions of the Royal Society of that year. He found that the standard yard of Elizabeth at the Exchequer marked 36.015 inches, Bird's parliamentary standard of 1758, 36.00023, and General Roy's scale 36.00036, on the scale of Troughton.

In the year 1818, Captain Henry Kater, one of the commissioners of the prince regent, with the same microscopic beam compass of Troughton, measured the same scale of General Roy, and found 39.4 inches on the latter to be equal to 39.40144 on the scale of Troughton. The difference between these two results is $\frac{105}{10000}$, or rather more than a hundredth part of an inch. Captain Kater, to account for it, supposes, that when Sir George Shuckburg made the comparison, the two scales were not at the same temperature: but Sir George Shuckburg, in his own account of his experiments, expressly mentions his leaving together another of the scales with that of Troughton, by which he measured it, 24 hours, that they might acquire the same temperature; and marks the state of the temperature (51.7) when he measured the scale of General Roy. Captain Kater states the thermometer, when he measured it, to have been at 70.

A difference equally striking has happened in the experiments made in France and England, to ascertain the relative proportions of the English foot and of the French metre. The result of numerous experiments, made in France under the direction of the National Institute, or Academy of Sciences, has been to announce the metre to be precisely equal to 39.3824 English inches. The result of Captain Kater's experiments, after numerous others under the direction of the Royal Society, is the declaration, that the French metre is equal to 39.3708 English inches. The difference is $\frac{116}{10000}$ of an inch, more than one hundredth part, and as near as possible the same as that of the experiments of Captain Kater and of Sir George Shuckburg Evelyn, upon the scales of Troughton and of General Roy.

A very interesting account of experiments made in this country by Mr. Hassler, to ascertain the length of the metre, is subjoined to this report, from which the mean length of four standard metres was found to be 39.38024797 English inches upon a scale of Troughton's, of equal perfection with that of Sir George Shuckburg Evelyn.

Again; in the year 1814, the Committee of the House of

Commons resolved, and, in 1815, the House itself enacted, that the length of a pendulum vibrating seconds, in the latitude of London, had been ascertained to be 39.13047 inches of Bird's parliamentary standard yard.

In the year 1818, Captain Kater reported, as the result of his experiments, that the length of the pendulum vibrating seconds *in vacuo* at the level of the sea, at the temperature of 62° of Fahrenheit, in latitude 51° 31′ 8″ 4‴ north (London), was 39.13842 inches of the same Bird's parliamentary standard yard.

The difference is $\frac{795}{10000}$, or a one hundred and twenty-sixth part of an each.

By assuming a mean average from all these experiments, and the yard of Elizabeth at the Exchequer (the standard from which all the long measures of the United States are derived), as the measure of comparison, we might be warranted in taking 39.38 English as the length of the French platina standard metre, and 39.14 as the length of the pendulum vibrating seconds in the latitude of London. And if the attempt at a minuter decimal fraction than that of the 100th part of an inch in the making of metallic measures, should terminate again in disappointment, it is nevertheless true, that, to obtain accuracy even to that extent, the microscopic beam compasses, and the micrometer marking subdivisions to the 25,000th part of an inch, are essential auxiliaries: for in this, as in all the energies, moral or physical, of man, the pursuit of absolute perfection is the only means of arriving at the nearest approximation to it, attainable by human power.

When the proportion shall be thus ascertained, by a concurrent agreement with France, the act might declare

that the foot measure of the United States is to the standard platina metre of France in such proportion that
39.3802 inches are equal to the metre, and that 472.5623
millimetres are equal to the foot. The proportion of the
troy and avoirdupois pounds to the kilogramme might be
ascertained with equal accuracy, and declared in like
manner. A platina metre and kilogramme, being exact
duplicates of those in the French national archives, should
then be deposited and preserved with the national standards of the United States.

It is not proposed that the standard yard-measure of
the United States should be made of platina; but that
it should be of the same metal as the yard of 1601, at the
Exchequer, from which it will be taken. The very extraordinary properties of platina, its unequalled specific
gravity, its infusibility, its durability, its powers of resistance against all the ordinary agents of destruction and
change, give it advantages and claims to employment as a
primary standard for weights and measures, and coins, to
which no other substance in nature has equal pretensions.
The standard metre and kilogramme of France are of that
metal. Should the fortunate period arrive when the improvement in the moral and political condition of man
will admit of the introduction of one universal standard
for the use of all mankind, it is hoped and believed that
the platina metre will be that measure. But, as the principle respectfully recommended in this report is that of
excluding all innovation or change, for the present, of
our existing weights and measures, it is with a view to
uniformity that the preference is given, for the choice of
a new standard, to the same metal of which that measure

consists which has been the standard of our forefathers from the first settlement of the English colonies, and is exactly coeval with them. It is not unimportant that the standards, to be transmitted to the several States of the Union, should be of the same metal as the national standards, of which they shall be copies. The changes of the atmosphere produce different degrees of expansion and contraction upon different metals; and, when a measure of brass or copper is to be taken from a measure of platina, the differences of their expansibility become subjects of calculation, upon data not yet ascertained to entire perfection. The selection of platina for the French kilogramme has been attended with the singular consequence, that the standard of the archives is not of the same *weight* as the standard for use. The latter is of brass; and the copies taken from it for the real purposes of life are of the same weight in the air, but not of the same weight as the platina standard, because that is the weight of the cubic decimetre of distilled water *in vacuo*. Whenever calculations of allowances for atmospheric changes in the different metals are introduced into the comparison of measures, estimates take the place of certainty; and different results proceed from different times, places, or persons. The very immutability of platina, therefore, makes it unsuitable for a practical standard of mutable things. Change, and not stability, is the uniform measure of change. Justice consists in estimating everything by the law of its nature: and, to illustrate this idea by applying it to moral relations, it may be observed, that, to bring mutable substances to the test of immutable standards, would be like charging disembodied spirits to pass sentence by the laws of their superior nature upon the frailties and infirmities of man.

The plan which is thus, in obedience to the injunction of both houses of Congress, submitted for their consideration, consists of two parts, the principles of which may be stated: 1. To fix the standard, with the partial uniformity of which it is susceptible, for the present, excluding all innovation. 2. To consult with foreign nations, for the future and ultimate establishment of universal and permanent uniformity. An apology is due to Congress for the length, as well as for the numerous imperfections, of this report. Embracing views, both theoretical and historical, essentially differ from those which have generally prevailed upon the subject to which it relates, they are presented with the diffidence due from all individual dissent encountering the opinions of revered authority. The resolutions of both houses opened a field of inquiry so comprehensive in its compass, and so abundant in its details, that it has been, notwithstanding the lapse of time since the resolution of the Senate, as yet but very inadequately explored. It was not deemed justifiable to defer longer the answer to the calls of both houses, even if their conclusion from it should be the propriety rather of further inquiry than of immediate action. In freely avowing the hope that the exalted purpose, first conceived by France, may be improved, perfected, and ultimately adopted by the United States, and by all other nations, equal freedom has been indulged in pointing out the errors and imperfections of that system, which have attended its origin, progress, and present condition. The same liberty has been taken with the theory and history of the English system, with the further attempt to show that the latter was, in its origin, a system of beauty, of symmetry, and of usefulness, little inferior to that of modern France.

The two parts of the plan submitted are presented distinctly from each other, to the end that either of them, should it separately obtain the concurrence of Congress, may be separately carried into execution. In relation to weights and measures throughout the Union, we possess already so near an approximation to uniformity of law, that little more is required of Congress for fixing the standard than to provide for the uniformity of fact, by procuring and distributing to the executives of the States and Territories positive national standards conformable to the law. If there be one conclusion more clear than another, deducible from all the history of mankind, it is the danger of hasty and inconsiderate legislation upon weights and measures. From this conviction, the result of all inquiry is, that, while all the existing systems of metrology are very imperfect, and susceptible of improvements involving in no small degree the virtue and happiness of future ages; while the impression of this truth is profoundly and almost universally felt by the wise and the powerful of the most enlightened nations of the globe; while the spirit of improvement is operating with an ardor, perseverance and zeal, honorable to the human character, it is yet certain, that, for the successful termination of all these labors, and the final accomplishment of the glorious object, permanent and universal uniformity, legislation is not alone compe-

tent. A concurrence of will is indispensable to give efficacy to the precepts of power. All trifling and partial
attempts of change in our existing system, it is hoped, will
be steadily discountenanced and rejected by Congress; not
only as unworthy of the high and solemn importance of
the subject, but as impracticable to the purpose of uniformity, and as inevitably tending to the reverse, to
increased diversity, to inextricable confusion. *Uniformity*
of weights and measures, permanent, universal uniformity,
adapted to the nature of things, to the physical organization, and to the moral improvement of man, would be a
blessing of such transcendent magnitude, that, if there
existed upon earth a combination of power and will adequate to accomplish the result by the energy of a single
act, the being who should exercise it would be among the
greatest of benefactors of the human race. But this stage
of human perfectibility is yet far remote. The glory of
the first attempt belongs to France. France first surveyed
the subject of weights and measures in all its extent and
all its compass. France first beheld it as involving the
interests, the comforts, and the morals, of all nations and
of all after ages. In forming her system, she acted as the
representative of the whole human race, present and to
come. She has established it by law within her own territories; and she has offered it as a benefaction to the acceptance of all other nations. That it is worthy of their
acceptance, is believed to be beyond a question. But
opinion is the queen of the world; and the final prevalence
of this system beyond the boundaries of France's power
must await the time when the example of its benefits, long

and practically enjoyed, shall acquire that ascendency over the opinions of other nations which gives motion to the springs and direction to the wheels of power.

Respectfully submitted,

JOHN QUINCY ADAMS.

DEPARTMENT OF STATE, *February* 22, 1821.

PART IV.

LECTURE OF SIR JOHN HERSCHEL.

THE attention of the public has of late been strongly drawn to the subject of a proposed alteration of our national system of weights and measures, by the attempt made during the last session of Parliament to carry through a bill, having for its object the abolition of our existing system in its entirety, and the introduction, in its place, of what is known as the "French Metrical System." The bill, it is true, was withdrawn after passing the second reading (by which the House, as is usually supposed, "*affirmed the principle of the measure*"), and it may therefore be reasonably presumed that it will be brought forward again in the next session, in the same or a modified form. As the discussion it received in the House seemed to be in no respect commensurate with the immense importance and sweeping nature of the change proposed, and with the exception of one or two rather cursory notices in *The Times*, excited a marvellously small amount of public interest pending its progress; it will not be amiss if, being called upon by the committee of the Leeds Astronomical Society for an exposition of some point of general interest in the form of a lecture or essay, to be read at one of their

13*

evening meetings, I select this for its subject; and endeavor to place before you the several conditions which any standard or typical unit of length which shall be assumed as the basis of a system of measures and weights intended to be national, and which may justly claim to be universal, ought to fulfil; and to compare with these conditions, in order to see how far they are fulfilled in fact, both our actual standard, the French metre now in use, and the length of the pendulum, which has been more than once proposed as a natural unit of length. And this I will endeavor to do in as elementary and familiar a way as shall be consistent with perfect correctness. Those of the present audience who are not already familiar with the subject will thus be better enabled to form an opinion as to the desirableness of the change actually proposed, or of any legislative change in our existing standard, and in our system of measures, weights, and coinage generally. And to such it will not be amiss to observe in the outset that, the subject being an exceedingly delicate and refined one, they must not be surprised at seeing very minute quantities and very small fractions treated as matters of much greater importance than they may have been accustomed to regard them.

2. The general subject of a national system of weights and measures, be it observed, divides itself into two very distinct and separate points of inquiry, viz., first, What is intrinsically the best and most available unit of linear measure to adopt as a basis; and, secondly, what system of numerical multiplication and aliquot subdivision of such unit for measures of length, and of its derivative units of area, of capacity, and of weight (for these all refer them-

selves naturally and easily to the unit of linear measure, or at least ought to do so) is most advantageous—either in a great mercantile community like our own, or for the great mass of mankind in the ordinary transactions of life. And it cannot be too strongly impressed, and too perseveringly borne in mind, that these two questions stand in no natural and necessary relation to each other, but are perfectly independent. We may resolve, with perfect logical consistency, either to toss aside our present system *in toto,* and adopt the metrical one in preference; or to retain our fundamental unit (the imperial foot or yard), and decimalize our system of denominations; or, lastly, by a slight, and, practically speaking, imperceptible change in our present standard, to bring it into conformity with our views of theoretical perfection (which, I shall show, may be done). We may, too, retaining all the convenience of our existing denominations (*so far as they are convenient*) superadd to them, by permissive legislation, the additional convenience of a decimal system for facility of calculation: relying on its holding its ground if really affording such facility, or working its way into general use, and ultimately driving out the old system, if found by the mass of the population to be practically preferable. This last is the course I would myself prefer, and I think it best to say so in the outset, lest those who may take a contrary view should imagine a foregone conclusion to be urged upon them under the semblance of free inquiry.

3. It is unnecessary, of course, to observe that, the measurement of length being required for almost every purpose of construction as well as for every intelligible statement of the sizes of material objects, the lengths of journeys, the

distances of places, etc.—renders indispensable the recognition, in every community, of some common standard, some well known and identifiable *unit*, by whose repetition great, and by whose aliquot subdivision small lengths, distances, sizes, etc., may be expressed in words and numbers. The common sense of mankind, moreover, would naturally point, in the selection of such unit, to some object of common occurrence, of moderate linear dimension, and of which individual exemplars differed but little, or, if possible, not at all in this respect; so that appeal might at once be made to such exemplar in case of a question arising as to the length of any object stated to contain a given number of such units or its aliquots. A very moderate experience would, however, suffice to convince anybody that among natural objects of the same kind, even those most common, perfect identity of length, of breadth, of thickness, any more than of weight, is never observed—even a close approach to it rarely—and a very close one extremely so. Still, with all drawbacks so arising on the adoption of a natural standard, the first rude demand for such a standard would be easily enough satisfied, and that in two ways, viz., 1st, by actually fixing upon some individual among all the existing objects of the sort selected, to the exclusion of others—or, 2dly, by the very natural, though somewhat more refined conception of an ideal medium, or *mean* among a very great multitude of such objects, such as might be regarded as neither unusually great nor unusually little ones of their kind.

4. Among objects of common occurrence, the human person, or some distinct member of it, would be most likely to claim the attention of mankind as affording a standard

of measure; if only for the very obvious reason that the relation of the sizes of material objects to that of man mainly determines his facility of handling, or otherwise applying them to human uses. Accordingly, the height of a full grown person, the length of his arm, his fore-arm (ulna or ell), his foot, his hand, his ordinary step, etc., would present, and is well known to have presented, itself among almost all communities of mankind to their choice for this purpose. And so, among all nations whose measures have been handed down to us, we find in speaking of the unit of length, some members of the human person designated. Thus, the bed of the gigantic king of Basan is related to have measured eight cubits in length " after the cubit (*i. e.*, the fore-arm) of a man." The height of Goliath the Philistine was "six cubits *and a span*." The bow of Pandarus, described by Homer, was formed of the horns of an ibex, which grew out sixteen *palms* (or hand-breadths) from his head. The Romans reckoned their distances by intervals of 1000 *paces* (*millia passuum*), whence our *name* for a mile, though differing widely in reality. If, however, we may judge from the great diversity in the actual lengths adopted under the common name of "a foot" as the standards of different nations, we shall see reason to believe that the typical foot selected was usually that of an individual—some chief, king, or high-priest, who could claim pre-eminence among them as a man *par excellence*, and who would seem to have been generally above the average stature. Thus we find the Roman foot equivalent to 11.6 of our inches; the English to 12; the Greek to 12.1; the French to 12.8; and the Egyptian or "Drusian" to 13.1—all of them (especially the two last) in excess of the real length of the foot of a well-proportioned man of medium stature (say 5ft. 10in.), which does not exceed $10\frac{3}{4}$, or at the most 11 inches.

5. Another class of objects, which, from the universality of their occurrence in vast numbers, and their general uniformity of dimension, would naturally occur as unit types, available for the measurement of small lengths, or for the small aliquots of a larger unit, has been found in the cereal grains of most common use, and of these, the barley-corn, and the rice-grain, have found the preference. Our inch, for instance, has been defined in an old statute (now repealed) as the length of three grains of barley, taken from the middle of the ear, and placed end to end. And in a somewhat similar manner have been derived from those cereals the smaller subdivisions of the Hebrews and Hindoos; while the larger have, in these, as in other nations, originated in parts of the human person.

6. It is very evident, however, that types of this kind admit of no precise and rigorous identification or inter-comparison. The medium stature of a man is very different in different countries. That of an adult French conscript, for instance, is (or at least was in 1817) 5 ft. 4in., as concluded from the measurement of 100,000 individuals, while the Belgian type, or mean adult stature, has been placed at 5ft. 7in. .8, and that of a Lancashire non-manufacturing laborer, as high as 5ft. $10\frac{3}{4}$in. So great a discordance as a result of local and secondary circumstances, is of course fatal to the pretensions of the human person as a natural type. So again of the cereals. The difference of soil, climate, and cultivation must produce, and does in fact produce, very great variety in the medium size of grain

grown in different countries, and in different years: so that, even supposing them to be measured by millions, the mean results would be found to differ too much for the object in view. And the same kind of objection holds good against having recourse to any kind of medium magnitude, among multitudes of objects of a like species which occur in nature. Such must, of necessity, be chosen among organic structures of the animal or vegetable kingdom (for among inorganic masses of whatever kind, nature presents no instance of a *mean* or typical magnitude, as distinct from the *average* of a number accidentally assembled, which may differ to any extent from an average similarly taken of an equal number elsewhere collected). And among the former classes of objects, even were it possible to assemble and measure them in sufficient numbers to afford a true *typical mean,* we should have no security for its identity in different ages and climates.

7. We are driven, then, in our choice of a *universal* standard, to the selection either of some individual object (if such there be), natural or artificial, imperishable in its nature, unsusceptible of variation by lapse of time or decay, and indestructible by accident—or else, to some ideal or resultant length or magnitude (if such there be), susceptible by its definition of being as it were translated into a material expression, and marked out as the result of some process which we are sure will, in all ages and places, reproduce the same identical result. And besides these qualities of invariability, indestructibility and identical reproducibility, it ought to possess some obvious claim to *general* acceptation as of common interest to all mankind, or at least to all the civilized portion of it: an interest

from which national partialities and rivalries should be altogether excluded.

8. The individual human type is at once excluded by these conditions. Supposing the foot of the most remarkable person who ever lived to be marked out on steel or adamant, it would be at the mercy of fire, earthquake, loss in political convulsions, and a hundred other forms of destruction or disappearance, without the possibility of re-appeal to the original form. Of human works, the most permanent, no doubt, and the most imposing as well as generally interesting and respected, are those mighty monumental structures which have been erected as if for the purpose of defying the powers of elementary change. Take the vastest of them—that to which appeal has been often made for this very purpose—the great pyramid of Cheops. When built it was 481 ft. in height, and the square area of its base was 764 ft. in the side. The height is now only 451 ft., and the side of the base only 746; and the sole means by which we are now enabled to determine the original height consists in a block of the exterior marble casing, which will in all probability disappear in the hands of "the curious" within the next century. Nature presents to us but one material *object* which combines all the requisites enumerated, and combines them all in perfection—viz., the globe itself that we inhabit. And in that globe we find only two naturally-defined lengths which unite the requisites of individuality to identify them under every change of human relations and even of geological revolutions and catastrophes, and of universality, so as to stand in the same relation to both hemispheres and to all meridians—viz., the earth's polar axis, and its equatorial

circumference. For the latter, the equatorial *diameter* might be more advantageously substituted: but that we have good reason to believe the equator to be not strictly circular, but in some degree elliptic, the proportion of its greatest and least diameters not being yet precisely known, though very much nearer to equality than that of the equatorial and polar diameters. This however would not prevent its *mean* equatorial diameter from being assumed in preference to its circumference, were not the polar axis, for very obvious reasons, preferable to both. Of the latter, and indeed of all three (thanks to the elaborate geodesical surveys which have been made within the century last elapsed), we possess a knowledge so precise as to render them perfectly available for our purpose.

9. Of lengths which exist not marked by the dimensions of any material object, but which are defined by the nature of things and by physical relations, and which are susceptible of exact determination and of being marked off on a scale, and so of becoming materialized for practical reference; there have been proposed only three which can be considered theoretically, and of these only one practically available. These are, 1st, the velocity of light or the space travelled over by light in some definite time (say the ten-millionth part of a second, which would give a modulus of about 100 feet); 2dly, the length of an undulation of a ray of light of some definite refrangibility—a length so minute as to require multiplication a million-fold to give a modular unit; and 3dly, the length of a pendulum vibrating seconds under certain definite and normal circumstances—or rather that of an ideal seconds-pendulum supposed to be placed at the extremity of the earth's polar

axis. To this is in effect equivalent, and derivable from it, as a mere arithmetical conclusion, the space fallen through by a heavy body on the same place by the earth's attraction in a second of time. The modulus so obtained is therefore a measure of the earth's total attractive power (independent of centrifugal force arising from its rotation), as that derived from the length of its diameter is of its total bulk, and equally unalterable and universal. As for the other two which depend on the nature of light, the difficulty and delicacy of the processes they would involve render all idea of resorting to either of them purely visionary.

10. The linear dimensions of the earth, then, on the one hand, and the linear measure of its attractive force embodied in the pendulum on the other, are the two, and, so far as we can see, the only two available sources of the invariable and universal standard length which we seek. And it is curious to observe that while the French, after considering both of them, threw aside the pendulum in favor of the metre (or ten-millionth of the meridian quadrant); the English, on the other hand, by the Act of Parliament in 1824, which repealed the old statute already alluded to (and so threw aside the principle of resorting to an organic type) did in effect, at that time, adopt the pendulum as their ultimate resort. For while that act declares that a certain metallic bar made by Bird in 1760, when at the temperature of 62° Fahr. should, without any further reference to its origin, be considered the standard yard of the British empire, it provided for its recovery and reproduction in case of the total destruction or loss of it and all its authentic copies and facsimiles, by a declaration that its length is 36 inches, *such* that 39.13929 of them are

equal to the length of a pendulum vibrating seconds *in vacuo* and at the sea-level, in the latitude of London. The report of the French commissioners also in 1798, which led to the enactment of the metrical system, is careful to state that in the event of the total loss or destruction of all material representatives of the metre its value would be easily recoverable from a numerically specified relation between its length and that of the pendulum vibrating seconds at Paris, which had been determined with great accuracy by Borda, one of the commissioners. So that, practically speaking, in the event of the total destruction, by political convulsions, of every authentic yard and metre (supposing any written record of our existing knowledge to survive them) the metre would have been recovered, not by the laborious and costly process of remeasuring the French meridian arc, but by the infinitely more summary one of a precise repetition of Borda's experiments and the exact re-application of all his corrections and reductions.

11. For the reproduction of the English yard, a similar repetition of those experiments *in* London which led to the adoption of the number 39.13929 in. as the measure of the pendulum would, in such an event, no doubt have been, at that epoch, resorted to; though in departure from the wording of the act, which speaks of a pendulum vibrating seconds, not *at* but *in the latitude of* London: a very different thing, as General Sabine has pointed out in his "*Account of Experiments to determine the figure of the Earth by means of a Pendulum vibrating Seconds in different latitudes.*" For the object would have been then, as it really was on the occasion of the actual destruction of the parliamentary standard in 1834, not to *produce* a theoreti-

cally *better*, but as far as possible to *reproduce* the same *identical* length by the most summary process; without undertaking circumnavigatory voyages, or entering on any theoretical discussion. The new act necessary for legalizing the standard so arising would probably have sanctioned this procedure, and we should have thenceforward had a standard of a purely local character, assuming for the fundamental basis the individual seconds pendulum *in* London.

12. This, however, is not now the case. On the destruction of the standard of 1760 by the burning of the Houses of Parliament, the new standard was constructed, not by any measurement of the length of the pendulum (for in the ten years elapsed since 1826 very grave doubts had been raised, or rather very serious sources of error pointed out in the processes used for the purpose on the former occasion)—but, by an assemblage and most careful comparison of all the scales and standards of any authority which could be got together—resulting in the production of one primary and a great many secondary standards, in all human probability absolutely identical with that destroyed. The act, moreover (of 1855), which constituted that one our legal yard, and named the others in a certain order as its successors in the event of its destruction or loss, omitted the clause identifying its length with any numerical multiple of the pendulum. In fact, then, our yard is a purely individual material object, multiplied and perpetuated by careful copying; and from which all reference to a natural origin is studiously excluded, as much as if it had dropped from the clouds. Apart, then, from the extraordinary pains taken in its construction, and from the sin-

gularly fortunate but at the same time purely accidental coincidence which I shall presently mention, it has no pretensions whatever to be regarded as a scientific unit.

13. Let us now consider the claim which the pendulum, in the abstract, as a measure of the earth's gravitation, can advance for its reception as a fundamental and universal standard of length (and here, incidentally, it may be remarked that, *as a length*, it is not more inconvenient than the metre, being within about a quarter of an inch the same).* One of the reasons assigned by the French *savans* for their rejection of it in favor of the metre, and, as would appear, the only one which weighed with them (for their other reason ostensibly advanced is a mere appeal to the political passions of the time), was the dependence of the length of the pendulum on the time of its vibration; as if the 86,400th part of a day, which we call a second of time, were not as definite and as invariable a quantity as the 100,000th part which, in their rage for decimalization, they proposed to call one; and as if they might not have fixed on a pendulum vibrating 100,000 times in a day (which would have given a very near approach to our yard). But their stumbling-block was the introduction of an extraneous element, *time*, at all, into the subject: as if the length of the day were not as much

---

* The metre has this inconvenience, as compared with the yard—that while the latter can be readily extemporized by a man of ordinary stature (and often is so in practice) by holding the end of a string or ribbon between the finger and thumb of one hand at the full length of the arm extended horizontally sideways, and marking the point which can be brought to touch the centre of the lips (facing full in front); the former is considerably too long to afford the same facility.

an invariable, universal, and physical element as the dimensions of the earth or its gravitation. But in this they seem to have overlooked the fact that their adoption of the quadrant of a meridian for the base of their system does really admit this extraneous element, time, into that system, though in a much more insidious way. For the total bulk or mean radius and the total mass or gravitating energy of the earth remaining the same, the ellipticity of its meridians, and therefore their absolute length, depends on the period of its rotation or the length of the day. The same objection, to be sure, if it be one, would equally apply to the adoption of the polar axis, or the equatorial diameter of the earth; and the only way to exclude all ideas of time and force from a metrical system, and render it *purely* metrical, *i. e.*, dependent on geometrical magnitude alone, would be to take for a fundamental unit the radius, diameter, or circumference of a sphere, or the side of a cube, equal in volume to that of the earth. And perhaps were a *tabula rasa* made; were the ground totally unoccupied and the whole matter to do over again, this would be as good a unit as could be proposed.

14. But the true objection to the choice of the pendulum for a universal unit of measure lies, not in any metaphysical and abstract considerations of this kind; but in the uncertainty which prevails, and must necessarily always prevail, as to the true length of that normal or ideal pendulum which shall stand equally related to the whole globe, and from which the mean length corresponding to any assigned latitude can be calculated: that is to say, the length of a pendulum which would swing seconds at the pole of the terrestrial spheroid—an uncertainty which, as

I shall proceed to show, must affect the result of every attempt to deduce it with the precision the subject requires from experiments made on the surface of our planet: however refined the methods employed, and however numerous and diversified the geographical stations at which they may be instituted.

15. In practice, the mean length of the polar or equatorial pendulum is concluded from an assemblage of the observations of the times of oscillation of one and the same invariable pendulum at a multitude of geographical stations in all accessible latitudes in both hemispheres: no two combinations agreeing in giving the same precise length, by reason of the local deviations of the intensity of gravity due to the nature of the soil, and the configuration of the ground immediately beneath and around the places of observation. Now, since the pendulum cannot be observed at sea, the whole sea-covered surface of the globe is of necessity excluded from furnishing its quota of observations to the final or mean conclusion. And the influence of this, it should be observed, is not self-compensating as that of local inequalities of mere density on land would be, but tells all in one direction. For water being, on the average, not more than one-third the weight of an equal bulk of land (such land as the earth's surface consists of) and only $\frac{2}{11}$ of the mean density of the globe, the force of gravity at the surface of the sea is less than at the sea-level on land by the attractive force of as much material taken at twice the specific gravity of water, or at $\frac{4}{11}$ths that of the globe, as would be required to raise the bottom to the surface. Supposing then the difficulty of observing the pendulum at sea overcome, and that the whole surface

of the globe were dotted over with stations of observation equally distributed over sea and land, from whose inter-comparison it were required to derive the mean co-efficient of terrestrial gravitation, or the mean length of the polar pendulum; it is evident that the sea-stations would everywhere conspire to give a less result than the land. According to Dr. Young (Phil. Trans., vol. cix., page 93) the attraction of an extensive flat mass of any thickness on a point in the middle of its surface is three times that of a sphere of the same materials having that thickness for its diameter. And from this it is very easy to conclude that, supposing the sea to have a mean depth of four miles (which seems not improbable) the mean defalcation of gravity at its surface, due to the deficiency of attracting matter, would be three times the attraction of a sphere four miles in diameter and $\frac{4}{11}$ths of the earth's mean density—that is by a simple calculation $\frac{1}{1833}$, or rather less than one 1800th part of the whole attraction of the earth—a fraction far too large, as well as far too uncertain in its amount either at any given spot or in general, not to vitiate irremediably any conclusion as to the ultimate result of the operation.

16. Similarly, if we look to the reductions to the sea-level necessary for stations in the interior of continents, we shall find that they depend, partly on the diminution of gravity due to the *height above* the sea-level, or to the increase of distance from the earth's centre, which always tells in *diminution* of gravity; and partly on the protuberant matter, be it mountain or elevated table-land immediately beneath and around the pendulum, which always tells in favor of *increased* gravitation. The former portion is rigorously calculable, and therefore need not trouble us,

but the latter is in an extreme degree uncertain in particular localities, and in a general estimate falls very short
of compensating for the sea-deficiency. For the mean
height of the European continent is only 1342 feet; of
Asia 2274; of North America 1496; and of South America 2302. The mean is 1840 feet, or rather more than a
third of a mile, which, on the same principle of reckoning,
would be equivalent to about $\frac{1}{15000}$th part only of the
total gravity, which has to be reduced to one-third of its
amount, or to $\frac{1}{45000}$th, inasmuch as the proportion of land
to water over the whole globe is only that of 51 to 146, or
about 1 to 3. This is the mean effect of the elevated matter to *increase* gravitation. That of mere *elevation* above
the sea-level to the height of $\frac{1}{3}$ of a mile (similarly reduced)
is, however, one 36,000th in the opposite direction, or to
diminish it—and the difference or one 180,000th of the
whole is effective *not to compensate but to add to* the sea-
deficiency.

17. To obtain the real length of the normal pendulum,
then, we must go out of our own globe, and ascertain the
true co-efficient of gravity from astronomical facts; and, as
the only one available for the purpose, compute the distance
fallen through by the moon in a second of time toward the
earth from a tangent to her orbit. This, it is evident, is independent of the influence of those local inequalities which
affect the pendulum measurements. But, on the other
hand, it must be remembered—1st, That our knowledge
of the distance in question depends on our previous knowledge of the moon's distance, which, in its turn, depends on
that of the earth's diameter, and therefore presupposes the
metre to be *accurately* known. For any *aliquot* error in

the metre will produce an equal *aliquot* error in the moon's distance estimated in metres, and therefore also in the linear deflection per second from the tangent to the orbit. 2d, That this linear deflection, or approach of the moon to the earth in one second of time, is the result of the joint attraction of the earth on the moon and of the moon on the earth, and is in effect the sum of the spaces fallen through by the moon toward their common centre of gravity, in virtue of the earth's attraction, and by the earth toward that point in virtue of the moon's. Now the mass of the moon is about one 88th part of that of the earth, so that one 88th part of the force that draws them together is due to the moon. By so much then must the space fallen through be diminished, to get that due to the earth's alone. Suppose, now, that the moon's mass assumed should be in error by a 50th part of its whole amount (and Laplace's estimate of it differs by as much from that at present received), and we shall find ourselves landed, from this cause of uncertainty alone, in an error to the extent of nearly one 4000th of the quantity sought.

18. Lastly, our knowledge of the moon's mass is mainly derived from its effect in producing the phenomenon of nutation, which it does through the medium of the earth's ellipticity, so that not only the dimensions, but the figure of the earth, are thus mixed up in our attempt to derive the length of the normal pendulum from the moon's motion.

19. I cannot but consider, then, that the uncertainty of the one mode of obtaining the length of the normal pendulum, and the non-independence of the other, unfit it for being received as the ultimate scientific basis of a universal

standard; whatever merit it may possess in an abstract and metaphysical point of view—and that the true and only practical use of the pendulum in relation to such a standard is the ready, cheap, and perfectly unobjectionable means its measurement, at a determinate spot and under defined circumstances, affords of recovering it when lost, by the recorded statement of its length in terms of such standard.

20. The causes of uncertainty which tell with such very appreciable effect on the local determination of the force of gravity by the pendulum, have little or no influence on the local curvature of the surface of equilibrium, and absolutely none on the measures of large arcs of the meridian. Suppose, for example, a sea of four miles in depth, and of great extent, to cover one part of the earth's surface. Its surface water will gravitate less by one 1800th part of its proper weight, owing to the deficiency of attracting matter below it; and the diminution of gravity growing less and less in descending (being proportional to the height of a particle above the bottom), the whole weight of the column of water vertically above a given spot will be diminished by one 3600th part of four miles, or one 900th of a mile, *i. e.*, about six feet of additional water, must be heaped on: a mere infinitesimal of the radius of curvature of its surface, which is that of the earth itself.

21. Let us now see how far the French metre, as it stands, fulfills the requirements of scientific and ideal perfection. It professes to be the 10,000,000th part of the quadrant of the meridian passing through France from Dunkirk to Formentera, and is therefore, scientifically speaking, a local and national, and not a universal measure. The earth's equator is not a perfect circle, but slightly elliptic, and the

meridians of places differing in longitude are therefore not
all of the same length. The difference, however, is so trifling (the ellipticity of its equator being not more than a
thirtieth part of that of its meridian) that to raise an objection against the practical reception of the metre, either
per se, or as a substitute for the yard, on this score, would
savor of hyper-criticism.   A more serious objection is the
choice made of the circumference of the meridional or
generating ellipse of the terrestrial spheroid in preference
to its axis of revolution. This is a blemish on the very
face of the system—a sin against geometrical simplicity.
Still, were the length of the metre as determined by the
French geometers rigorously exact, or correct within limits
which the much more extensive measurements of meridian
arcs since made elsewhere than in France have proved to
be attainable, this would be only a matter of regret, and
could hardly, of itself, be drawn into an argument for its
rejection. But this is far from being really the case. The
metre, as represented by the material standard adopted as
its representative, is too short by a sensible and measurable
quantity, though one which certainly might be easily corrected. To show this it will be necessary to enter into
some detail.

22. In effect, that standard is declared, in the Annuary
of the Bureau des Longitudes, to be equal to 39.37079
British imperial standard inches. The quadrant of the
French meridian then ought, if this be correct, to be 393,-
707,900 such inches, or 32,808,992 feet. And by whatever
aliquot part of its whole length the true quadrant exceeds
this, by that same aliquot of *its* length is the metre, so
stated, erroneous.

23. Mr. Airy, by a combination of the whole series of meridian arcs whose measures had been obtained in every part of the globe in 1830, was led to conclude for the value of the minor or polar axis of the terrestrial spheroid, 41.707,620 feet; while the late Professor Bessel, pursuing a course similar in its general principle—that is to say, using all the measured arcs, great and small, in combination one with another, and taking the most probable mean among the (necessarily) discordant results, obtained by combining them two and two—arrived at a value very slightly different, viz., 41,707,314 feet. The mean of these gives, as the result of this mode of procedure, 41,707,467.

24. Quite recently, M. Schubert, in a very elaborate memoir which appears as part of the 1st vol., 7th series, of the Memoirs of the Petersburg Academy, has pointed out the inconvenience, and necessarily discordant results which the combination by pairs of a multitude of small arcs, each of itself insufficient to afford any precise measure of the ellipticity, affords; and assigned his reasons for restricting the inquiry in the first instance into the length of the polar axis, as an element unique in itself, and common to all the meridians: deducing it separately from each of the most extensive arcs, the Russian, the Indian, and the French, each taken independently;—comparing the three values so obtained, and thence concluding the final result. In this manner he obtains the following three values of the axis, viz.:—

From the Russian arc (of 25° 20′ in extent) 41,711,019.2 feet.
  "   Indian " (of 21° 21′ ") 41,712,534.2 feet.
  "   French " (of 12° 22′ ") 41,697,496.4 feet.

In concluding for these a mean, or final value, M. Schubert however, arbitrarily, and as I think quite indefensibly, rejects altogether the result of the French arc, and assigns to the Russian double the weight of the Indian; a mode of procedure in which he will find, I presume, few to agree with him. A much fairer, indeed the only fair way to treat them, is obviously to ascribe to each of the separate results in taking the mean, a weight proportional to the total extent of the arc, and this gives for the length of the axis 41,708,710.0 feet. Comparing then the final results of the modes of procedure we find,

> From the former...................... 41,707,467 feet
> And from the latter.................... 41,708,710 "

which differ only by 1243 feet, or less than $\frac{1}{4}$ of a mile—so that their mean or 41,708,088.5 f. is in all probability within a furlong, or one part in 64,000 of the truth.

25. From each of the great arcs of Russia and India, M. Schubert then obtains a separate value of the equatorial or the larger axis of the elliptic meridian to which it belongs; and by a similar treatment of the arc of Peru, which lying under the equator, is especially favorable for the purpose, he obtains a third value of the equatorial diameter. The three diameters of the equatorial ellipse thus obtained, with the angles they include at the centre (which are the differences of longitude of the respective meridians, and which are as favorably arranged for the purpose as the nature of the case seems to admit), suffice for the determination of the major and minor axis of the equator, regarded as an ellipse, and the longitudes in which they lie, viz. :—

Axis major = 41,854,800 feet, in long. 38° 44′ E. from Paris (one end falling about half-way between Mount

Kenia and the east coast of Africa, the other in the middle of the Pacific Ocean).

Axis minor = 41,850,007 feet, in long. 128° 44′ E. from Paris (one end falling on Waygiou, one of the Molucca Islands, and the other at the mouth of the Amazon River), giving an ellipticity of one 8880th, or about one-thirtieth part of that of the meridians as already stated.

26. The figure of the equator, and its dimensions thus obtained, the exact equatorial diameter corresponding to any given longitude is easily calculated. And by comparing this with the polar axis, the precise ellipticity of the meridian for that longitude may be computed. And executing this computation for Paris, M. Schubert finds $\frac{1}{288}$ for the ellipticity of the French meridian.

27. With these data, viz., a polar axis of 41,708,088 feet, and an ellipticity of $\frac{1}{288}$ which certainly may lay claim to greater precision than anything previously obtained, I shall now proceed to calculate the true length of the quadrant of the French meridian, for which purpose the following very simple and convenient formula may be used,* viz. :—

$$Q = \frac{\pi}{4} A (1 + 2m + 9m^2 + 38m^3)$$

in which Q represents the length of the quadrant required,

---

* For the present purpose it is necessary to carry out the calculation to the cube of ellipticity—but in cases where the *square* of that fraction may be neglected, the following simple rule for finding the circumference of an ellipse is worth remembering. On the longer axis of the ellipse describe a circle, and between this and the ellipse, describe a small circle having its centre in the prolongation of the minor axis, and touching the ellipse externally, and the circumscribed circle internally. The circumference of this small circle is the difference between those of the ellipse and of the larger or circumscribing circle.

A that of the polar axis, π the circumference of a circle whose diameter is 1, and m, *one fourth part* of the fraction expressing the ellipticity, or in this case $\frac{1}{1184}$.

Executing the calculation the result is.....32,813,000 feet,
Subtract 10,000,000 metres = 32,808,992
Remain, excess......... 4,008

for the excess of the true quadrant over the assumed as the basis of the metrical system, that is to say, one 8194 aliquot part of the whole, or one 208th of an inch on the whole metre, which is therefore the quantity by which the French standard is actually too short.

28. It must not be denied that this is a very wonderful approximation, and in the highest degree creditable to the science, skill, and devotion of the French astronomers and geometricians who carried on their operations under every difficulty, and at the hazard of their lives in the midst of the greatest political convulsion of modern times. And adopted as it is over a large portion of Europe; were the question an open one what standard a new nation, unprovided with one, unfettered by usages of any sort, and in the absence of any knowledge of the existence of the British yard, should select; there could be no hesitation as to its adoption (with that very slight correction above pointed out—which would in no way interfere with its practical use—a correction which the French themselves might, under such circumstances, consent to adopt). But the question now arising is quite another thing, viz., whether we are to throw overboard an existing, established, and, so to speak, ingrained system—adopt the metre as it stands, for our standard—adopt moreover its decimal sub-

divisions, and carry out the change into all its train of consequences; to the rejection of our entire system of weights, measures, and coins. If we adopt the metre we cannot stop short of this. It would be a standing reproach and anomaly—a change for changing sake. The change, if we make it, must be complete and thorough. And this in the face of the fact that England is beyond all question the nation whose commercial relations, both internal and external, are the greatest in the world, and that the British system of measures is received and used, not only throughout the whole British empire (for the Indian " Hath," or revenue standard, is defined by law to be 18 British imperial inches), but throughout the whole North American continent, and (so far as the measure of length is concerned) also throughout the Russian empire; the standard unit of which, the Sagene, is declared by an imperial ukase to contain *exactly* seven British imperial feet, and the Archine and Vershock precise fractions of the Sagene. Taking commerce, population, and area of soil then into account, there would seem to be far better reason for our continental neighbors to conform to *our* linear unit could it advance the same, or a better *à priori* claim, than for the move to come from our side. (I say nothing at present of decimalization.)

29. Let us see then how this part of the matter stands. Taking the polar axis of the earth as the best unit of dimensions which the terrestrial spheroid affords (a better *à priori* unit* than that of the metrical system), we have

---

* A writer in *Quesneville's Moniteur Scientifique*, No. 163, v. 736, argues that *itinerary* measures ought to be based on the *circumference*

seen that it consists of 41,708,088 imperial feet—which, reduced to inches, is 500,497,056 imperial inches. Now this differs only by 2944 inches, or by 82 yards from 500,500,000 (five hundred million and five hundred thousand) such inches—and this would be the whole error on a length of 8000 miles which would arise from the adoption of this precise round number of inches for its length, or from making the inch, so defined, our fundamental unit of length. Suppose, then, that any length were proposed in English measure, and we desire to know what decimal fraction such length were of the earth's axis. We have only to express it in inches and decimals, and from the number so stated take off its thousandth part (a calculation involving only the writing down the number twice over, removing the figures of the under line three places to the right and subtracting), and the thing is done, and *vice versâ*.* Suppose now the same length stated in French

---

of the globe and not on its *axis*—by reason that the decimal principle of subdivision, if carried out, would apply to the decimal graduation of the quadrant—*adding* that " the greatest advantage of the French system is in reality its decimal division,"—but *forgetting* to add that the decimal division of the quadrant *was* introduced in France, but *was abandoned by common consent even in France*, and can never be reintroduced. In the "*Mondes*" (Suppl. 38, p. 616) the same argument is advanced, and the same answer applies.

* Strictly speaking for the conversion and reconversion we should *subtract* one 999th and *add* one 1000th. But the difference is only one part in a million which can never be of the slightest importance. *Per contra* the conversion of the metre according to the process here stated leads to a result which, though exact in parts of the *French* meridian, is erroneous in parts of the *mean terrestrial* meridian by a considerably larger proportional part, and *this* is what we really want to know.

metres, and we would ascertain what decimal fraction it is of a quadrant of the French meridian. The number of metres assigned must be divided by 8194 either by a long division sum or by the use of a table, before the proper number to be subtracted can be found. Which then is the shorter process? and which, both scientifically and practically, the preferable unit?

30. If we are to legislate at all on the subject then, the enactment ought to be to increase our present standard yard (and of course all its multiples and submultiples) by one precise thousandth part of their present lengths, and we should then be in possession of a system of linear measure the purest and most ideally perfect imaginable. The change, so far as relates to any practical transaction, commercial, engineering, or architectural, would be absolutely unfelt, as there is no contract for work even on the largest scale, and no question of ordinary mercantile profit or loss, in which one *per mille* in measure or in coin would create the smallest difficulty. Neither could it be doubted that our example would be very speedily followed both in America and Russia, so soon as the reason of the thing and the trifling amount of the change came to be understood. And even without legislation the relation between the proposed new or *geometrical* measure and the imperial ones is so simple and striking—fixing itself so easily in the memory, and the conversion from one to the other so ready, that, *were there no other reason*, it might almost be questioned whether it would be worth while to make the change.

31. But there *is* another reason, and I think a decisive one. Hitherto I have said nothing about our weights and

measures of capacity. Now, as they stand at present nothing can be more clumsy and awkward than the numerical connection between the seand our unit of length. A grain is defined as the weight of distilled water, so that, 252.724 of such grains at the freezing temperature, or 252.46 at that of 62° Fahr. which is the standard temperature of our imperial yard, shall fill a cubic inch. Of such grains, so defined, the pound contains 7000, the ounce 437½, and the gallon of water at 62°, 70,000. According to this system, the cubic foot of water at our standard temperature weighs 997.145 oz., falling short of 1000 oz. by very nearly 3 oz. However tempting this approximation might appear, still, in the absence of any more cogent reason, the commissioners who recommended our system of weights and measures legalized in 1824 forbore to recommend such a change in the ounce (about 1½ grain) as would have brought it about; though the rule that a cubic foot of water weighs 1000 ounces, is still handed down as a rough and ready way of converting cubic measure into weight. But were we to adopt the *geometrical* instead of the present imperial standard—the linear foot being increased by one thousandth, the cubic foot would be increased by three times that aliquot, or would become 1.003 times our present cubic foot—and so would make up just the deficient three ounces, or at least so very nearly that a legislative change in the ounce, increasing it by only one part in 8000, or by one 18th part of a grain, would bring everything into decimal coincidence, by making the ounce and the cubic foot the links of connection between weights and measures instead of the grain and the cubic inch, as at present. As regards our measures of *capacity*, the connection would be equally consecutive, as a decimal one, between the cubic foot and the half-pint, which for the purpose in view, ought to have a distinct name (such as a "*tumbler*," or a "*rummer*," or a "*beaker*")—and which would contain exactly one 100th part of a cubic foot,—with whatever liquid or solid matter it might be filled. And thus the change which would place our system of linear measure on a perfectly faultless basis, would at the same time rescue our weights and measures of capacity from their present utter confusion, and secure that other advantage, second only in importance to the former, of connecting them decimally with that system on a regular, intelligible, and easily-remembered principle; and *that* by an alteration practically imperceptible in both cases, and interfering with no one of our usages or denominations.

32. On the subject of decimalization, it will be gathered from what I have said that I would make any decimalized denominations which anybody might agree to buy, sell, or contract by, permissive. There seems to be a doubt whether such is now the case, and if so the law should, I think, be altered. But I would leave untouched all our present *denominations* and their relations to the standard —and the only new measure I would legalize would be a "module" (or some other name *at present unoccupied*) of 50 geometrical inches, being the ten millionth of the polar axis, or its half, the "geometrical cubit" of 25 such inches —leaving its use quite voluntary.

COLLINGWOOD, *Sept.* 30, 1863.

ADDENDUM.

33. Since the foregoing remarks were written my attention has been called by the Astronomer Royal to a very elaborate memoir by Captain Clarke, in vol. **xxix.** of the Memoirs of the Royal Astronomical Society, whose conclusions, though differing from those of M. Schubert in some particulars (as in making the equator more elliptic), yet, so far as the present subject is concerned, tend in the same direction, and that, as regards the aliquot error of the metre, even more strongly.

34. Captain Clarke assigns for the three axes of the earth the following values :—

Polar axis	41,707,536 feet
Or in inches	500,490,432
Longer equatorial axis	41,852,970 feet
Shorter do. do.	41,842,354 "

Longitude of the vertex of the longer axis $= 13°\ 58'\ 30''$ east—(or $11°\ 35'\ 15''$ E. of Paris) whence it is easy to conclude as follows :—

Diameter of equator in the longitude of Paris....41,852,695 feet.
Ellipticity of the Paris meridian...................$\frac{1}{2927\cdot7}$ say $\frac{1}{292}$

35. Calculating now the quadrant from this ellipticity, and from Captain Clarke's polar axis, we find it 32,814,116 feet, which exceeds ten million metres by 5124 feet, being in excess of that above found (4008) by 1116 feet; and corresponding to an aliquot error of one part in 6404, or on the metre itself to one 163d part of an inch. The aliquot error in our " geometrical yard " is also somewhat increased by the adoption of this polar axis, viz., to one part in 52,310, or to about one 1453d part of an inch on the yard.

36. As this memoir of Captain Clarke contains by far the most complete and comprehensive discussion which the subject of the earth's figure has yet received, and must be held as the ultimatum of what scientific calculation is as yet enabled to exhibit as its true dimensions and form— this conclusion will of course be considered to supersede that arrived at in the foregoing pages.

COLLINGWOOD, *Oct.* 11, 1863.

P.S.—Some slight subsequent corrections made by Capt. Clarke in his calculations, founded on data quite recently published, make the polar axis approximate *still more nearly* to 500,500,000 inches.

ORTHOGRAPHY AND READING.

NATIONAL SERIES

OF

READERS AND SPELLERS.

BY PARKER & WATSON.

The National Primer	$ 25
National First Reader	38
National Second Reader	63
National Third Reader	1 00
National Fourth Reader	1 50
National Fifth Reader	1 88
National Elementary Speller	25
National Pronouncing Speller	50
Independent Third Reader	75
Independent Fourth Reader	1 00
Independent Fifth Reader	1 25

The salient features of these works which have combined to render them so popular may be briefly recapitulated as follows:

1. THE WORD-BUILDING SYSTEM.—This famous progressive method for young children originated and was copyrighted with these books. It constitutes a process with which the beginner with *words* of one letter is gradually introduced to additional lists formed by prefixing or affixing single letters, and is thus led almost insensibly to the mastery of the more difficult constructions. This is justly regarded as one of the most striking modern improvements in methods of teaching.

2. TREATMENT OF PRONUNCIATION.—The wants of the youngest scholars in this department are not overlooked. It may be said that from the first lesson the student by this method need never be at a loss for a prompt and accurate rendering of every word encountered.

3. ARTICULATION AND ORTHOEPY are recognized as of primary importance.

1

(over.)

ORTHOGRAPHY AND READING—Continued.

4. PUNCTUATION is inculcated by a series of interesting *reading lessons*, the simple perusal of which suffices to fix its principles indelibly upon the mind.

5. ELOCUTION. Each of the higher Readers (3d, 4th and 5th) contains elaborate, scholarly, and thoroughly practical treatises on elocution. This feature alone has secured for the series many of its warmest friends.

6. THE SELECTIONS are the crowning glory of the series. Without exception it may be said that no volumes of the same size and character contain a collection so diversified, judicious, and artistic as this. It embraces the choicest gems of English literature, so arranged as to afford the reader ample exercise in every department of style. So acceptable has the taste of the authors in this department proved, not only to the educational public but to the reading community at large, that thousands of copies of the Fourth and Fifth Readers have found their way into public and private libraries throughout the country, where they are in constant use as manuals of literature, for reference as well as perusal.

7. ARRANGEMENT. The exercises are so arranged as to present constantly alternating practice in the different styles of composition, while observing a definite plan of progression or gradation throughout the whole. In the higher books the articles are placed in formal sections and classified topically, thus concentrating the interest and inculcating a principle of association likely to prove valuable in subsequent general reading.

8. NOTES AND BIOGRAPHICAL SKETCHES. These are full and adequate to every want. The biographical sketches present in pleasing style the history of every author laid under contribution.

9. ILLUSTRATIONS. These are plentiful, almost profuse, and of the highest character of art. They are found in every volume of the series as far as and including the Third Reader.

10. THE GRADATION is perfect. Each volume overlaps its companion preceding or following in the series, so that the scholar, in passing from one to another, is barely conscious, save by the presence of the new book, of the transition.

11. THE PRICE is reasonable. The books were not *trimmed* to the minimum of size in order that the publishers might be able to denominate them " the cheapest in the market," but were made *large enough* to cover and suffice for the grade indicated by the respective numbers. Thus the child is not compelled to go over his First Reader twice, or be driven into the Second before he is prepared for it. The competent teachers who compiled the series made each volume just what it should be, leaving it for their brethren who should use the books to decide what constitutes true *cheapness.* A glance over the books will satisfy any one that the same amount of matter is nowhere furnished at a price more reasonable. Besides which another consideration enters into the question of relative economy, namely, the

12. BINDING. By the use of a material and process known only to themselves, in common with all the publications of this house, the National Readers are warranted to outlast any with which they may be compared—the ratio of relative durability being in their favor as two to one.

Parker & Watson's National Series of Readers
TESTIMONIALS.

From Hon. T. A. Parker, State Sup't of Public Instruction, Missouri.

By authority of law it becomes my duty to recommend a list of Text-books for use in the Public Schools of Missouri. I deem it necessary to approve a list of books which will secure to the youth of the State a *uniform, cheap, and practical* course of study, and after careful examination have selected the following. The National Readers and Spellers, *Monteith & McNally's Geographies, Peck's Ganot's Natural Philosophy, Jarvis' Physiology and Health, &c., &c.*

From Sam'l P. Bates, LL.D., Asst. Supt. Public Schools of Pennsylvania.

I find that your series of Parker & Watson's National Readers are going into use in all our leading Normal Schools. They are unquestionably ahead of any thing yet published.

From A. J. Haile, Prin. Hebrew Educational Institute, Memphis, Tennessee.

I take great pleasure in bearing testimony to the superior merits of Parker & Watson's Series of " National Readers."

From Prof. F. S. Jewell, of the New York State Normal School.

It gives me pleasure to find in the National Series of School-Readers ample room for commendation. From a brief examination, I am led to believe that we have none equal to them. I hope they will prove as popular as they are excellent.

From Moses T. Brown, Superintendent Public Schools, Toledo, Ohio.

The different Series of other authors were critically examined by our Board of Education and myself, and the decision was unanimous in favor of the National Series. Our teachers are delighted with the books, and none more so than our primary teachers. *I consider the Series better adapted to our graded school system than any other now before the public.*

From Wm. B. Ames, Superintendent of Schools, Morris, Connecticut.

They are well adapted to all degrees of scholarship—one lesson prepares the mind of the pupil for the next in consecutive order, from book to book—till the highest order of English composition is attained in the Fifth Reader.

From John S. Hart, Prin. N. J. State Normal School.

I approve of Parker & Watson's Readers highly. The selections are judicious, the arrangement good, and the books well made mechanically. We have adopted the 3d, 4th, and 5th of the Series in this school.

From R. P. Deckard, President Ewing College, La Grange, Texas.

I think the National Series of Readers the best I have seen.

Extracts from Report made to the California State Teachers' Association.

The Committee, in presenting to this Convention the Series of Readers by Parker & Watson, would state that, regarded as a whole, we would give our unqualified support to them in preference to all others.

From B. J. Young, Superintendent Schools, Shelbyville, Illinois.

The National Readers have been selected for use in the public schools of this city, and are giving very excellent satisfaction. During ten years' experience in teaching, I have found no books so well adapted to secure rapid and thorough progress.

From the Wilmington (N. C.) Daily Herald.

The National Series has attained probably a higher reputation than any other complete series of School-Books in existence.

☞ For further testimony of a similar character, see special circular, or current numbers of the Educational Bulletin.

The National Readers and Spellers.

THEIR RECORD.

These books have been adopted by the School Boards, or official authority, of the following important States, cities, and towns—in most cases for exclusive use.

The State of Minnesota.

The State of Missouri.

The State of Alabama.

The State of North Carolina.

The State of Louisiana.

New York.
New York City.
Brooklyn.
Buffalo.
Albany.
Rochester.
Troy.
Syracuse.
Elmira.
&c., &c.

Pennsylvania.
Reading.
Lancaster.
Erie.
Scranton.
Carlisle.
Carbondale.
Meadville.
Schuylkill Haven.
Williamsport.
Norristown.
Bellefonte.
Altoona.
&c., &c.

New Jersey.
Newark.
Jersey City.
Paterson.
Trenton.
Camden.
Elizabeth.
New Brunswick.
Phillipsburg.
Orange.
&c., &c.

Delaware.
Wilmington.

D. C.
Washington.

Illinois.
Chicago.
Peoria.
Alton.
Springfield.
Aurora.
Galesburg.
Rockford.
Rock Island.
&c., &c.

Wisconsin.
Milwaukee.
Fond du Lac.
Oshkosh.
Janesville.
Racine.
Watertown.
Sheboygan.
La Crosse.
Waukesha.
Kenosha.
&c., &c.

Michigan.
Grand Rapids.
Kalamazoo.
Adrian.
Jackson.
Monroe.
Lansing.
&c., &c.

Ohio.
Toledo.
Sandusky.
Conneaut.
Chardon.
Hudson.
Canton.
Salem.
&c., &c.

Indiana.
New Albany.
Fort Wayne.
Lafayette.
Madison.
Logansport.
Indianapolis.

Iowa.
Davenport.
Burlington.
Muscatine.
Mount Pleasant.
&c.

Nebraska.
Brownsville.
Lincoln.
&c.

Oregon.
Portland.
Salem.
&c.

Virginia.
Richmond.
Norfolk.
Petersburg.
Lynchburg.
&c.

South Carolina.
Columbia.
Charleston.

Georgia.
Savannah.

Louisiana.
New Orleans.

Tennessee.
Memphis

The *Educational Bulletin* records periodically all new points gained.

SCHOOL-ROOM CARDS,

To Accompany the National Readers.

Eureka Alphabet Tablet *1 50

Presents the alphabet upon the Word Method System, by which the child will learn the alphabet in nine days, and make no small progress in reading and spelling in the same time.

National School Tablets, 10 Nos. *8 00

Embrace reading and conversational exercises, object and moral lessons, form, color, &c. A complete set of these large and elegantly illustrated Cards will embellish the school-room more than any other article of furniture.

READING.

Fowle's Bible Reader $1 00

The narrative portions of the Bible, chronologically and topically arranged, judiciously combined with selections from the Psalms, Proverbs, and other portions which inculcate important moral lessons or the great truths of Christianity. The embarrassment and difficulty of reading the Bible itself, by course, as a class exercise, are obviated, and its use made feasible, by this means.

North Carolina First Reader 40
North Carolina Second Reader 65
North Carolina Third Reader 1 00

Prepared expressly for the schools of this State, by C. H. Wiley, Superintendent of Common Schools, and F. M. Hubbard, Professor of Literature in the State University.

Parker's Rhetorical Reader 1 00

Designed to familiarize Readers with the pauses and other marks in general use, and lead them to the practice of modulation and inflection of the voice.

Introductory Lessons in Reading and Elocution 75

Of similar character to the foregoing, for less advanced classes.

High School Literature 1 50

Admirable selections from a long list of the world's best writers, for exercise in reading, oratory, and composition. Speeches, dialogues, and model letters represent the latter department.

ORTHOGRAPHY.

SMITH'S SERIES

Supplies a speller for every class in graded schools, and comprises the most complete and excellent treatise on English Orthography and its companion branches extant.

1. Smith's Little Speller . . . .$ 20

First Round in the Ladder of Learning.

2. Smith's Juvenile Definer 45

Lessons composed of familiar words grouped with reference to similar signification or use, and correctly spelled, accented, and defined.

3. Smith's Grammar-School Speller 50

Familiar words, grouped with reference to the sameness of sound of syllables differently spelled. Also definitions, complete rules for spelling and formation of derivatives, and exercises in false orthography.

4. Smith's Speller and Definer's Manual . 90

A complete *School Dictionary* containing 14,000 words, with various other useful matter in the way of Rules and Exercises.

5. Smith's Hand-Book of Etymology . . . 1 25

The first and only Etymology to recognize the *Anglo-Saxon our mother tongue;* containing also full lists of derivatives from the Latin, Greek, Gaelic, Swedish, Norman, &c., &c : being, in fact, a complete etymology of the language for schools.

Sherwood's Writing Speller 15
Sherwood's Speller and Definer 15
Sherwood's Speller and Pronouncer . . . 15

The Writing Speller consists of properly ruled and numbered blanks to receive the words dictated by the teacher, with space for remarks and corrections. The other volumes may be used for the dictation or ordinary class exercises.

Price's English Speller *15

A complete spelling-book for all grades, containing more matter than "Webster," manufactured in superior style, and sold at a lower price—consequently the cheapest speller extant.

Northend's Dictation Exercises 63

Embracing valuable information on a thousand topics, communicated in such a manner as at once to relieve the exercise of spelling of its usual tedium, and combine it with instruction of a general character calculated to profit and amuse.

Wright's Analytical Orthography 25

This standard work is popular, because it teaches the elementary sounds in a plain and philosophical manner, and presents orthography and orthoepy in an easy, uniform system of analysis or parsing.

Fowle's False Orthography 45

Exercises for correction.

Page's Normal Chart *3 75

The elementary sounds of the language for the school-room walls.

ENGLISH GRAMMAR.

CLARK'S DIAGRAM SYSTEM.

Clark's First Lessons in Grammar . . . 45
Clark's English Grammar · . · 1 00
Clark's Key to English Grammar 75
Clark's Analysis of the English Language · 60
Clark's Grammatical Chart · *3 75

The theory and practice of teaching grammar in American schools is meeting with a thorough revolution from the use of this system. While the old methods offer proficiency to the pupil only after much weary plodding and dull memorizing, this affords from the inception the advantage of *practical Object Teaching*, addressing the eye by means of illustrative figures; furnishes association to the memory, its most powerful aid, and diverts the pupil by taxing his ingenuity. Teachers who are using Clark's Grammar uniformly testify that they and their pupils find it the most interesting study of the school course.

Like all great and radical improvements, the system naturally met at first with much unreasonable opposition. It has not only outlived the greater part of this opposition, but finds many of its warmest admirers among those who could not at first tolerate so radical an innovation. All it wants is an impartial trial, to convince the most skeptical of its merit. No one who has fairly and intelligently tested it in the school-room has ever been known to go back to the old method. A great success is already established, and it is easy to prophecy that the day is not far distant when it will be the *only system of teaching English Grammar*. As the SYSTEM is copyrighted, no other text-books can appropriate this obvious and great improvement.

Welch's Analysis of the English Sentence · 1 25

Remarkable for its new and simple classification, its method of treating connectives, its explanations of the idioms and constructive laws of the language, &c.

ETYMOLOGY.

Smith's Complete Etymology, · . · . . . 1 25

Containing the Anglo-Saxon, French, Dutch, German, Welsh, Danish, Gothic, Swedish, Gaelic, Italian, Latin, and Greek Roots, and the English words derived therefrom accurately spelled, accented, and defined.

The Topical Lexicon, · . . · . · . . . 1 75

This work is a School Dictionary, an Etymology, a compilation of synonyms, and a manual of general information. It differs from the ordinary lexicon in being arranged by topics instead of the letters of the alphabet, thus realizing the apparent paradox of a "Readable Dictionary." An unusually valuable school-book.

Clark's Diagram English Grammar.

TESTIMONIALS.

From J. A. T. DURNIN, Principal Dubuque R. C. Academy, Iowa.

In my opinion, it is well calculated by its system of analysis to develop those rational faculties which in the old systems were rather left to develop themselves, while the memory was overtaxed, and the pupils discouraged.

From B. A. COX, School Commissioner, Warren County, Illinois.

I have examined 150 teachers in the last year, and those having studied or taught Clark's System have universally stood fifty per cent. better examinations than those having studied other authors.

From M. H. B. BURKET, Principal Masonic Institute, Georgetown, Tennessee.

I traveled two years amusing myself in instructing (exclusively) Grammar classes with Clark's system. The first class I instructed fifty days, but found that this was more time than was required to impart a theoretical knowledge of the science. During the two years thereafter I instructed classes only *thirty* days each. Invariably I proposed that unless I prepared my classes for a more thorough, minute, and accurate knowledge of English Grammar than that obtained from the ordinary books and in the ordinary way in from one to two years, I would make no charge. I never failed in a solitary case to far exceed the hopes of my classes, and made money and character rapidly as an instructor.

From A. B. DOUGLASS, School Commissioner, Delaware County, New York.

I have never known a class pursue the study of it under a *live* teacher, that has not succeeded; I have never known it to have an opponent in an educated teacher who had *thoroughly* investigated it; I have never known an *ignorant* teacher to examine it; I have never known a teacher who has used it, to try any other.

From J. A. DODOR, Teacher and Lecturer on English Grammar, Kentucky.

We are tempted to assert that it foretells the dawn of a brighter age to our mother-tongue. Both pupil and teacher can fare sumptuously upon its contents, however highly they may have prized the manuals into which they may have been initiated. and by which their expressions have been moulded.

From W. T. CHAPMAN, Superintendent Public Schools, Wellington, Ohio.

I regard Clark's System of Grammar the best published. For teaching the analysis of the English Language, it surpasses any I ever used.

From F. S. LYON, Principal South Norwalk Union School, Connecticut.

During ten years' experience in teaching, I have used six different authors on the subject of English Grammar. I am fully convinced that Clark's Grammar is better calculated to make thorough grammarians than any other that I have seen.

From CATALOGUE OF ROHRER'S COMMERCIAL COLLEGE, St. Louis, Missouri.

We do not hesitate to assert, without fear of successful contradiction, that a better knowledge of the English language can be obtained by this system in six weeks than by the old methods in as many months.

From A. PICKETT, President of the State Teachers' Association, Wisconsin.

A thorough experiment in the use of many approved authors upon the subject of English Grammar has convinced me of the superiority of Clark. When the pupil has completed the course, he is left upon a foundation of *principle*, and not upon the *dictum* of the author.

From GEO. F. McFARLAND, Prin. McAllisterville Academy, Juniata Co., Penn.

At the first examination of public-school teachers by the county superintendent, when one of our student teachers commenced analyzing a sentence according to Clark, the superintendent listened in mute astonishment until he had finished, then asked what that meant, and finally, with a very knowing look, said such work wouldn't do here, and asked the applicant to parse the sentence right, and gave the lowest certificates to all who barely mentioned Clark. Afterwards, I presented him with a copy, and the next fall he permitted it to be partially used, while the third or last fall, he openly commended the system, and appointed three of my best teachers to explain it at the two Institutes and one County Convention held since September.

☞ For further testimony of equal force, see the Publishers' Special Circular, or current numbers of the Educational Bulletin.

GEOGRAPHY.

NATIONAL GEOGRAPHICAL SYSTEM.

THE SERIES.

I. Monteith's First Lessons in Geography, . . . $ 85
II. Monteith's New Manual of Geography, . . . 1 10
III. McNally's System of Geography, ... 2 00

INTERMEDIATE OR ALTERNATE VOLUMES.

1*. Monteith's Introduction to the Manual, . . . 63
2*. Monteith's Physical and Intermediate Geography, . 1 75

ACCESSORIES.

Monteith's Wall Maps (per set) *20 00
Monteith's Manual of Map-Drawing, ... 25
Monteith's Map-Drawing and Object-Lessons, . . 75
Monteith's Map-Drawing Scale, ... *25

1. **PRACTICAL OBJECT TEACHING.** The infant scholar is first introduced
to *a picture* whence he may derive notions of the shape of the earth, the phenomena
of day and night, the distribution of land and water, and the great natural divisions,
which mere words would fail entirely to convey to the untutored mind. Other pictures follow on the same plan, and the child's mind is called upon to grasp no idea
without the aid of a pictorial illustration. Carried on to the higher books, this system
culminates in No. 2*, where such matters as climates, ocean currents, the winds, peculiarities of the earth's crust, clouds and rain, are pictorially explained and rendered
apparent to the most obtuse. The illustrations used for this purpose belong to the
highest grade of art.

2. **CLEAR, BEAUTIFUL, AND CORRECT MAPS.** In the lower numbers
the maps avoid unnecessary detail. while respectively progressive, and affording the
pupil new matter for acquisition each time he approaches in the constantly enlarging
circle the point of coincidence with previous lessons in the more elementary books.
In No. 2*, the maps embrace many new and striking features. One of the most
effective of these is the new plan for displaying on each map the relative sizes of
countries not represented, thus obviating much confusion which has arisen from the
necessity of presenting maps in the same atlas drawn on different scales. The maps
of No. 3 have long been celebrated for their superior beauty and completeness. This
is the only school-book in which the attempt to make a *complete* atlas *also clear and
distinct*, has been successful. The map *coloring* throughout the series is also noticeable. Delicate and subdued tints take the place of the startling glare of inharmonious
colors which too frequently in such treatises dazzle the eyes, distract the attention,
and serve to overwhelm the names of towns and the natural features of the landscape.

GEOGRAPHY—Continued

3. THE VARIETY OF MAP EXERCISE Starting each time from a different basis, the pupil in many instances approaches the same fact no less than *six times,* thus indelibly impressing it upon his memory. At the same time this system is not allowed to become wearisome—the extent of exercise on each subject being graduated by its relative importance or difficulty of acquisition.

4. THE CHARACTER AND ARRANGEMENT OF THE DESCRIPTIVE TEXT. The cream of the science has been carefully culled, unimportant matter rejected, elaboration avoided, and a brief and concise manner of presentation cultivated. The orderly consideration of topics has contributed greatly to simplicity. Due attention is paid to the facts in history and astronomy which are inseparably connected with, and important to the proper understanding of geography—and *such only* are admitted on any terms. In a word, the National System teaches geography as a science, pure, simple, and exhaustive.

5. ALWAYS UP TO THE TIMES. The authors of these books, editorially speaking, never sleep. No change occurs in the boundaries of countries, or of counties, no new discovery is made, or railroad built, that is not at once noted and recorded, and the next edition of each volume carries to every school-room the new order of things.

6. SUPERIOR GRADATION. This is the only series which furnishes an available volume for every possible class in graded schools. It is not contemplated that a pupil must necessarily go through every volume in succession to attain proficiency. On the contrary, *two* will suffice, but *three* are advised; and if the course will admit, the whole series should be pursued. At all events, the books are at hand for selection, and every teacher, of every grade, can find among them one *exactly suited* to his class. The best combination for those who wish to abridge the course consists of Nos. 1, 2, and 3, or where children are somewhat advanced in other studies when they commence geography, Nos. 1*, 2, and 3. Where but *two* books are admissible, Nos. 1* and 2*, or Nos. 2 and 3, are recommended.

7. FORM OF THE VOLUMES AND MECHANICAL EXECUTION The maps and text are no longer unnaturally divorced in accordance with the time-honored practice of making text-books on this subject as inconvenient and expensive as possible. On the contrary, all map questions are to be found on the page opposite the map itself, and each book is complete in one volume. The mechanical execution is unrivalled. Paper and printing are everything that could be desired, and the binding is—A. S. Barnes and Company's.

8. MAP-DRAWING. In 1869 the system of Map-Drawing devised by Professor JEROME ALLEN was secured *exclusively* for this series. It derives its claim to originality and usefulness from the introduction of *a fixed unit of measurement* applicable to every Map. The principles being so few, simple and comprehensive, the subject of Map-Drawing is relieved of all practical difficulty. (In Nos. 2, 2*, and 3, and published separately.)

9. ANALOGOUS OUTLINES. At the same time with Map-Drawing was also introduced (in No. 2), a new and ingenious variety of Object Lessons, consisting of a comparison of the outlines of countries with familiar objects pictorially represented.

Monteith & McNally's National Geographies.

CRITICAL OPINIONS.

From R. A. ADAMS, Member of Board of Education, New York.

I have found, by examination of the Book of Supply of our Board, that considerably the largest number of any series now used in our public schools is the National, by Monteith and McNally.

From BRO. PATRICK, Chief Provincial of the Vast Educational Society of the
CHRISTIAN BROTHERS in the United States.

Having been convinced for some time past that the series of Geographies in use in our schools were not giving satisfaction, and came far short of meeting our most reasonable expectations, I have felt it my imperative duty to examine into this matter, and see if a remedy could not be found.

Copies of the different Geographies published in this country have been placed at our command for examination. On account of other pressing duties we have not been able to give as much time to the investigation of all these different series as we could have desired; yet we have found enough to convince us that there are many others better than those we are now using; but we cheerfully give our most decided preference, above all others, to the National Series, by Monteith & McNally.

Their easy gradation, their thoroughly practical and independent character, their comprehensive completeness as a full and accurate system, the wise discrimination shown in the selection of the subject matter, the beautiful and copious illustrations, the neat cut type, the general execution of the works, and *other excellencies*, will commend them to the friends of education everywhere.

From the "HOME MONTHLY," Nashville, Tenn.

MONTEITH'S AND MCNALLY'S GEOGRAPHIES.—Geography is so closely connected with Astronomy, History, Ethnology, and Geology, that it is difficult to define its limits in the construction of a text-book. If the author confines himself strictly to a description of the earth's surface, his book will be dry, meager, and unintelligible to a child. If, on the other hand, he attempts to give information on the cognate sciences, he enters a boundless field, and may wander too far. It seems to us that the authors of the series before us have hit on the happy medium between too much and too little. *The First Lessons*, by applying the system of object-teaching, renders the subject so attractive that a child, just able to read, may become deeply interested in it. The second book of the course enlarges the view, but still keeps to the maps and simple descriptions. Then, in the third book, we have Geography combined with History and Astronomy. A general view of the solar system is presented, so that the pupil may understand the earth's position on the map of the heavens. The first part of the fourth book treats of Physical Geography, and contains a vast amount of knowledge compressed into a small space. It is made bright and attractive by beautiful pictures and suggestive illustrations, on the principle of object-teaching. The maps in the second part of this volume are remarkably clear, and the map exercises are copious and judicious. In the fifth and last volume of the series, the whole subject is reviewed and systematized. This is strictly a Geography. Its maps are beautifully engraved and clearly printed. The map exercises are full and comprehensive. In all these books the maps, questions and descriptions are given in the same volume. In most geographies there are too many details and minute descriptions—more than any child out of purgatory ought to be required to learn. The power of memory is overstrained; there is confusion—no clearly defined idea is formed in the child's mind. But in these books, in brief, pointed descriptions, and constant use of bright, accurate maps, the whole subject is photographed on the mind.

11

The National System of Geography,

By Monteith & McNally.

ITS RECORD.

These popular text-books have been adopted, by official authority, for the schools of the following States, cities, and associations—in most cases for exclusive and uniform use.

STATES.

California.
Missouri.
Alabama.
Tennessee.

Vermont.
Iowa.
Louisiana.

Minnesota.
North Carolina.
Kansas.

CITIES.

New York City.
Brooklyn.
New Orleans.
Buffalo.
Richmond.
Jersey City.
Hartford.
Worcester.

Louisville.
Newark.
Milwaukee.
Charleston.
Rochester.
Mobile.
Syracuse.
Memphis.

Nashville.
Utica.
Wilmington.
Trenton.
Norfolk.
Norwich.
Lockport.
Dubuque.

Portland.
Savannah.
Indianapolis.
Springfield.
Wheeling.
Toledo.
Bridgeport.
St. Paul.

ASSOCIATIONS.

The Society of the Christian Brothers, representing 40,000 pupils.
The Franciscan Brothers, 8,000 pupils.
American Missionary Society, 50,000 pupils.

Monteith's Physical & Intermediate Geography.

This is the most recently published of the Geographical Series, and as might have been anticipated, was very warmly received.

TESTIMONIALS IN BRIEF.

The more I examine the better I am pleased.—J. T. Goodnow, *State Supt. Kans.*
Has no superior as a text-book.—E. J. Thompson, *Supt. Fillmore Co., Minn.*
Brief, clear, suggestive, and admirably adapted.—E. Conant, *Prin. Vt. Normal.*
It is a gem of a book.—E. A. Strong, *Supt. Public Schools, Grand Rapids, Mich.*
The best adapted we have seen.—O. Faville, *State Supt., Iowa.*
A book that has long been needed.—A. J. Kingman, *Supt. McHenry Co., Ill.*
Prepared with labor, care, and well adapted.—C. B. Halstead, *Supt. Newburg, N. Y.*
The best Geography ever published.—J Hutchison, *Prin. Boys' Sch. Jefferson, La.*
I like it very much.—A. J. Craig, *State Superintendent, Wisconsin.*
Cannot fail to awaken a new interest.—*Vermont School Journal.*   [*Coll., Va.*
A new field cultivated with great success.—T. C. Johnson, *Pres, Randolph Macon.*
Contains more common sense than any other.—J. Angear, *Prin. Madison Ac. Iowa.*

MATHEMATICS.

DAVIES' NATIONAL COURSE.

ARITHMETIC.

SLATED

1. Davies' Primary Arithmetic . . . .$ 25
2. Davies' Intellectual Arithmetic 40
3. Davies' Elements of Written Arithmetic . . . 50 $ 65
4. Davies' Practical Arithmetic 1 00   1 10
   Key to Practical Arithmetic *1 00
5. Davies' University Arithmetic. 1 40   1 55
   Key to University Arithmetic *1 40

ALGEBRA.

1. Davies' New Elementary Algebra 1 25   1 40
   Key to Elementary Algebra *1 25
2. Davies' University Algebra 1 60   1 75
   Key to University Algebra *1 60
3. Davies' Bourdon's Algebra 2 25   2 45
   Key to Bourdon's Algebra *2 25

GEOMETRY.

1. Davies' Elementary Geometry and Trigonometry 1 40   1 55
2. Davies' Legendre's Geometry 2 25   2 45
3. Davies' Analytical Geometry and Calculus . . 2 50   2 70
4. Davies' Descriptive Geometry 2 75   3 00

MENSURATION.

1. Davies' Practical Mathematics and Mensuration 1 40   1 55
2. Davies' Surveying and Navigation 2 50   2 70
3. Davies' Shades, Shadows, and Perspective . . 3 75   4 00

MATHEMATICAL SCIENCE.

Davies' Grammar of Arithmetic . . . .* 50
Davies' Outlines of Mathematical Science *1 00
Davies' Logic and Utility of Mathematics *1 50
Davies & Peck's Dictionary of Mathematics *5 00

DAVIES' NATIONAL COURSE of MATHEMATICS.

ITS RECORD.

In claiming for this series the first place among American text-books, of whatever class, the Publishers appeal to the magnificent record which its volumes have earned during the *thirty-five years* of Dr. Charles Davies' mathematical labors. The unremitting exertions of a life-time have placed *the modern series* on the same proud eminence among competitors that each of its predecessors has successively enjoyed in a course of constantly improved editions, now rounded to their perfect fruition—for it seems indeed that this science is susceptible of no further demonstration.

During the period alluded to, many authors and editors in this department have started into public notice, and by borrowing ideas and processes original with Dr. Davies, have enjoyed a brief popularity, but are now almost unknown. Many of the series of to-day, built upon a similar basis, and described as "modern books," are destined to a similar fate ; while the most far-seeing eye will find it difficult to fix the time, on the basis of any data afforded by their past history, when these books will cease to increase and prosper, and fix a still firmer hold on the affection of every educated American.

One cause of this unparalleled popularity is found in the fact that the enterprise of the author did not cease with the original completion of his books. Always a practical teacher, he has incorporated in his text-books from time to time the advantages of every improvement in methods of teaching, and every advance in science. During all the years in which he has been laboring, he constantly submitted his own theories and those of others to the practical test of the class-room—approving, rejecting, or modifying them as the experience thus obtained might suggest. In this way he has been able to produce an almost perfect series of class-books, in which every department of mathematics has received minute and exhaustive attention.

Nor has he yet retired from the field. Still in the prime of life, and enjoying a ripe experience which no other living mathematician or teacher can emulate, his pen is ever ready to carry on the good work, as the progress of science may demand. Witness his recent exposition of the "Metric System," which received the official endorsement of Congress, by its Committee on Uniform Weights and Measures.

Davies' System is the acknowledged National Standard for the United States, for the following reasons :—

1st. It is the basis of instruction in the great national schools at West Point and Annapolis.

2d. It has received the *quasi* endorsement of the National Congress.

3d. It is exclusively used in the public schools of the National Capital.

4th. The officials of the Government use it as authority in all cases involving mathematical questions.

5th. Our great soldiers and sailors commanding the national armies and navies were educated in this system. So have been a majority of eminent scientists in this country. All these refer to "Davies" as authority.

6th A larger number of American citizens have received their education from this than from any other series.

7th. The series has a larger circulation throughout the whole country than any other, being *extensively used in every State in the Union.*

MATHEMATICS—Continued.

ARITHMETICAL EXAMPLES.

Reuck's Examples in Denominate Numbers $ 50
Reuck's Examples in Arithmetic 1 00

These volumes differ from the ordinary arithmetic in their peculiarly *practical* character. They are composed mainly of examples, and afford the most severe and thorough discipline for the mind. While a book which should contain a complete treatise of theory and practice would be too cumbersome for every-day use, the insufficiency of *practical* examples has been a source of complaint.

HIGHER MATHEMATICS.

Church's Elements of Calculus 2 50
Church's Analytical Geometry 2 50
Church's Descriptive Geometry, with Shades,
Shadows, and Perspective 4 00

These volumes constitute the "West Point Course" in their several departments.

Courtenay's Elements of Calculus 3 00

A work especially popular at the South.

Hackley's Trigonometry 3 00

With applications to navigation and surveying, nautical and practical geometry and geodesy, and logarithmic, trigonometrical, and nautical tables.

SLATED ARITHMETICS.

The Publishers have the pleasure to announce that they have perfected arrangements with the proprietor of Jocelyn's patent for Slated Books, whereby the "National Series of School Books" will enjoy the exclusive use of this remarkable and valuable invention. It consists of the application of an artificially slated surface to the inner cover of a book, with flap of the same opening outward, so that students may refer to the book and use the slate at one and the same time, and as though the slate were detached. When folded up, the slate preserves examples and memoranda till needed. The material used is as durable as the stone slate. The additional cost of books thus improved is trifling.

THE METRIC SYSTEM.

Resolution of the Committee of the House of Representatives on a "Uniform System of Coinage, Weights, and Measures."

Be it Resolved, That Professor Charles Davies, LL.D., of the State of New York, be requested to confer with superintendents of public instruction, and teachers of schools, and others interested in a reform of the present incongruous system, and by lectures and addresses, to promote its general introduction and use.

The official version of the Metric System, as prepared by Dr. Davies, may be found in the Written, Practical, and University Arithmetics of the Mathematical Series, and is also published separately, price postpaid, *five cents.*

Davies' National Course of Mathematics.

TESTIMONIALS.

From L. VAN BOKKELEN, *State Superintendent Public Instruction, Maryland.*

The series of Arithmetics edited by Prof. Davies, and published by your firm, have been used for many years in the schools of several counties, and the city of Baltimore, and have been approved by teachers and commissioners.

Under the law of 1865, establishing a uniform system of Free Public Schools, these Arithmetics were unanimously adopted by the State Board of Education, after a careful examination, and are now used in all the Public Schools of Maryland.

These facts evidence the high opinion entertained by the School Authorities of the value of the series theoretically and practically.

From HORACE WEBSTER, *President of the College of New York.*

The undersigned has examined, with care and thought, several volumes of Davies' Mathematics, and is of the opinion that, as a whole, it is the most complete and best course for Academic and Collegiate instruction with which he is acquainted.

From DAVID N. CAMP, *State Superintendent of Common Schools, Connecticut.*

I have examined Davies' Series of Arithmetics with some care. The language is clear and precise; each principle is thoroughly analyzed, and the whole so arranged as to facilitate the work of instruction. Having observed the satisfaction and success with which the different books have been used by eminent teachers, it gives me pleasure to commend them to others.

From J. O. WILSON, *Chairman Committee on Text-Books, Washington, D. C.*

I consider Davies' Arithmetics decidedly superior to any other series, and in this opinion I am sustained, I believe, by the entire Board of Education and Corps of Teachers in this city, where they have been used for several years past.

From JOHN L. CAMPBELL, *Professor of Mathematics, Wabash College, Inatana.*

A proper combination of abstract reasoning and practical illustration is the chief excellence in Prof. Davies' Mathematical works. I prefer his Arithmetics, Algebras, Geometry, and Trigonometry to all others now in use, and cordially recommend them to all who desire the advancement of sound learning.

From MAJOR J. H. WHITTLESEY, *Government Inspector of Military Schools.*

Be assured I regard the works of Professor Davies, with which I am acquainted, as by far the best text-books in print on the subjects which they treat. I shall certainly encourage their adoption wherever a word from me may be of any avail.

From T. McC. BALLANTINE, *Professor Mathematics, Cumberland College, Kentucky.*

I have long taught Prof. Davies' Course of Mathematics, and I continue to like their working.

From JOHN McLEAN BELL, B. A., *Principal of Lower Canada College.*

I have used Davies' Arithmetical and Mathematical Series as text-books in the schools under my charge for the last six years. These I have found of great efficacy in exciting, invigorating, and concentrating the intellectual faculties of the young.

Each treatise serves as an introduction to the next higher, by the similarity of its reasonings and methods; and the student is carried forward, by easy and gradual steps, over the whole field of mathematical inquiry, and that, too, in a *shorter* time than is usually occupied in mastering a single department. I sincerely and heartily recommend them to the attention of my fellow-teachers in Canada.

From D. W. STEELE, *Prin. Philekoian Academy, Cold Springs, Texas.*

I have used Davies' Arithmetics till I know them nearly by heart. A better series of school-books never were published. I have recommended them until they are now used in all this region of country.

A large mass of similar "Opinions" may be obtained by addressing the publishers for special circular for Davies' Mathematics. New recommendations are published in current numbers of the *Educational Bulletin.*

HISTORY.

Monteith's Youth's History, ...$ 75

A History of the United States for beginners. It is arranged upon the catechetical plan, with illustrative maps and engravings, review questions, dates in parentheses (that their study may be optional with the younger class of learners), and interesting Biographical Sketches of all persons who have been prominently identified with the history of our country.

Willard's United States, Sch. ed., $1 40. Un. ed. 2 25
Do.   do.   University edition, . 2 25

The plan of this standard work is chronologically exhibited in front of the title-page; the Maps and Sketches are found useful assistants to the memory, and dates, usually so difficult to remember, are so systematically arranged as in a great degree to obviate the difficulty. Candor, impartiality, and accuracy, are the distinguishing features of the narrative portion.

Willard's Universal History, ... 2 25

The most valuable features of the "United States" are reproduced in this. The peculiarities of the work are its great conciseness and the prominence given to the chronological order of events. The margin marks each successive era with great distinctness, so that the pupil retains not only the event but its time, and thus fixes the order of history firmly and usefully in his mind. Mrs. Willard's books are constantly revised, and at all times written up to embrace important historical events of recent date.

Berard's History of England, ... 1 75

By an authoress well known for the success of her History of the United States. The social life of the English people is felicitously interwoven, as in fact, with the civil and military transactions of the realm.

Ricord's History of Rome, ... 1 60

Possesses the charm of an attractive romance. The Fables with which this history abounds are introduced in such a way as not to deceive the inexperienced, while adding materially to the value of the work as a reliable index to the character and institutions, as well as the history of the Roman people.

Hanna's Bible History, ... 1 25

The only compendium of Bible narrative which affords a connected and chronological view of the important events there recorded, divested of all superfluous detail.

Summary of History, Complete 60
American History, $0 40. French and Eng. Hist. 35

A well proportioned outline of leading events, condensing the substance of the more extensive text-book in common use into a series of statements so brief, that every word may be committed to memory, and yet so comprehensive that it presents an accurate though general view of the whole continuous life of nations.

Marsh's Ecclesiastical History, ... 2 00
Questions to ditto, ... 75

Affording the History of the Church in all ages, with accounts of the pagan world during Biblical periods, and the character, rise, and progress of all Religions, as well as the various sects of the worshipers of Christ. The work is entirely non-sectarian, though strictly catholic.

PENMANSHIP.

Beers' System of Progressive Penmanship.
Per dozen $2 00

This "round hand" system of Penmanship in twelve numbers, commends itself by its simplicity and thoroughness. The first four numbers are primary books. Nos. 5 to 7, advanced books for boys. Nos. 8 to 10, advanced books for girls. Nos. 11 and 12, ornamental penmanship. These books are printed from steel plates (engraved by McLees), and are unexcelled in mechanical execution. Large quantities are annually sold.

Beers' Slated Copy Slips, per set *50

All beginners should practice, for a few weeks, slate exercises, familiarizing them with the form of the letters, the motions of the hand and arm, &c., &c. These copy slips, 32 in number, supply all the copies found in a complete series of writing-books, at a trifling cost.

Payson, Dunton & Scribner's Copy-B'ks. P. doz., *2 25

The National System of Penmanship, in three distinct series—(1) Common School Series, comprising the first six numbers; (2) Business Series, Nos. 8, 11, and 12; (3) Ladies' Series, Nos. 7, 9, and 10.

Fulton & Eastman's Chirographic Charts, *3 75

To embellish the school room walls, and furnish class exercise in the elements of Penmanship.

Payson's Copy-Book Cover, per hundred .*3 00

Protects every page except the one in use, and furnishes "lines" with proper slope for the penman, under. Patented.

National Steel Pens, Card with all kinds . . . *15

Pronounced by competent judges the perfection of American-made pens, and superior to any foreign article.

SCHOOL SERIES.		Index Pen, per gross 75	
School Pen, per gross, . .$ 60		BUSINESS SERIES.	
Academic Pen, do . . 63		Albata Pen, per gross, . . 40	
Fine Pointed Pen, per gross 70		Bank Pen, do . . 70	
POPULAR SERIES.		Empire Pen, do . . 70	
Capitol Pen, per gross, . . 1 00		Commercial Pen, per gross . 60	
do do pr. box of 2 doz. 25		Express Pen, do . 75	
Bullion Pen (imit. gold) pr. gr. 75		Falcon Pen, do . 70	
Ladies' Pen do 63		Elastic Pen, do . 75	

Stimpson's Scientific Steel Pen, per gross .*2 00

One forward and two backward arches, ensuring great strength, well-balanced elasticity, evenness of point, and smoothness of execution. One gross in twelve contains a Scientific Gold Pen.

Stimpson's Ink-Retaining Holder, per doz. .*2 00

A simple apparatus, whic.. does not get out of order, withholds at a single dip as much ink as the pen would otherwise realize from a dozen trips to the inkstand, which it supplies with moderate and easy flow.

Stimpson's Gold Pen, $3 00; with Ink Retainer *4 50
Stimpson's Penman's Card, ...* 50

One dozen Steel Pens (assorted points) and Patent Ink-retaining Pen holder.

BOOK-KEEPING

Smith & Martin's Book-keeping · · · · · $1 25
Blanks to ditto · · · · · · · · · · · · · *60

This work is by a practical teacher and a practical book-keeper. It is of a thoroughly popular class, and will be welcomed by every one who loves to see theory and practice combined in an easy, concise, and methodical form.

The Single Entry portion is well adapted to supply a want felt in nearly all other treatises, which seem to be prepared mainly for the use of wholesale merchants, leaving retailers, mechanics, farmers, &c., who transact the greater portion of the business of the country, without a guide. The work is also commended, on this account, for general use in Young Ladies' Seminaries, where a thorough grounding in the simpler form of accounts will be invaluable to the future housekeepers of the nation.

The treatise on Double Entry Book-keeping combines all the advantages of the most recent methods, with the utmost simplicity of application, thus affording the pupil all the advantages of actual experience in the counting-house, and giving a clear comprehension of the entire subject through a judicious course of mercantile transactions.

The shape of the book is such that the transactions can be presented as in actual practice; and the simplified form of Blanks, three in number, adds greatly to the ease experienced in acquiring the science.

DRAWING.

The Little Artist's Portfolio · · · · · · · · *50

25 Drawing Cards (progressive patterns), 25 Blanks, and a fine Artist's Pencil, all in one neat envelope.

Clark's Elements of Drawing · · · · · · *1 00

Containing full instructions, with appropriate designs and copies for a complete course in this graceful art, from the first rudiments of outline to the finished sketches of landscape and scenery.

Fowle's Linear and Perspective Drawing · 60

For the cultivation of the eye and hand, with copious illustrations and directions, which will enable the unskilled teacher to learn the art himself while instructing his pupils.

Monk's Drawing Books—Six Numbers, per set *2 25

A series of progressive Drawing Books, presenting copy and blank on opposite pages. The copies are fac-similes of the best imported lithographs, the originals of which cost from 50 cents to $1.50 *each* in the print-stores. Each book contains *eleven* large patterns. No. 1.—Elementary studies; No. 2.—Studies of Foliage; No. 3.—Landscapes; No. 4.—Animals, I.; No. 5.—Animals, II.; No. 6.—Marine Views, &c.

Ripley's Map Drawing · · · · · · · · · · 1 25

One of the most efficient aids to the acquirement of a knowledge of geography is the practice of map drawing. It is useful for the same reason that the best exercise in orthography is the *writing* of difficult words. Sight comes to the aid of hearing, and a double impression is produced upon the memory. Knowledge becomes less mechanical and more intuitive. The student who has sketched the outlines of a country, and dotted the important places, is little likely to forget either. The impression produced may be compared to that of a traveler who has been over the ground, while more comprehensive and accurate in detail.

NATURAL SCIENCE.

FAMILIAR SCIENCE

Norton & Porter's First Book of Science, · $1 75

By eminent Professors of Yale College. Contains the principles of Natural Philosophy, Astronomy, Chemistry, Physiology, and Geology. Arranged on the Catechetical plan for primary classes and beginners.

Chambers' Treasury of Knowledge, · · · · 1 25

Progressive lessons upon—*first*, common things which lie most immediately around us, and first attract the attention of the young mind; *second*, common objects from the Mineral, Animal, and Vegetable kingdoms, manufactured articles, and miscellaneous substances; *third*, a systematic view of Nature under the various sciences. May be used as a Reader or Text-Book.

NATURAL PHILOSOPHY.

Norton's First Book in Natural Philosophy, 1 00

By Prof. Norton, of Yale College. Designed for beginners; profusely illustrated, and arranged on the Catechetical plan.

Peck's Ganot's Course of Nat. Philosophy, 1 75

The standard text-book of France, Americanized and popularized by Prof. Peck, of Columbia College. The most magnificent system of illustration ever adopted in an American school-book is here found. For intermediate classes.

Peck's Elements of Mechanics, · · · · · · 2 25

A suitable introduction to Bartlett's higher treatises on Mechanical Philosophy, and adequate in itself for a complete academical course.

Bartlett's Synthetic Mechanics, · · · · · 3 75
Bartlett's Analytical Mechanics, · · · · · 5 50
Bartlett's Acoustics and Optics, · · · · · 3 00

A system of Collegiate Philosophy, by Prof. Bartlett, of West Point Military Academy.

Steele's 14 Weeks Course in Philosophy, · 1 50

GEOLOGY.

Page's Elements of Geology, · · · · · · · 1 25

A volume of Chambers' Educational Course. Practical, simple, and eminently calculated to make the study interesting.

Emmon's Manual of Geology, · · · · · · · 1 25

The first Geologist of the country has here produced a work worthy of his reputation. The plan of presenting the subject is an obvious improvement on older methods. The department of Palæontology receives especial attention.

Peck's Ganot's Popular Physics.

TESTIMONIALS.

From PROF. ALONZO COLLIN, *Cornell College, Iowa.*

I am pleased with it. I have decided to introduce it as a text-book.

From H. F. JOHNSON, *President Madison College, Sharon, Miss.*

I am pleased with Peck's Ganot, and think it a magnificent book.

From PROF. EDWARD BROOKS, *Pennsylvania State Normal School.*

So eminent are its merits, that it will be introduced as the text-book upon elementary physics in this institution.

From H. H. LOCKWOOD, *Professor Natural Philosophy U. S. Naval Academy.*

I am so pleased with it that I will probably add it to a course of lectures given to the midshipmen of this school on physics.

From GEO. S. MACKIE, *Professor Natural History University of Nashville, Tenn.*

I have decided on the introduction of Peck's Ganot's Philosophy, as I am satisfied that it is the best book for the purposes of my pupils that I have seen, combining simplicity of explanation with elegance of illustration.

From W. S. McRAE, *Superintendent Vevay Public Schools, Indiana.*

Having carefully examined a number of text-books on natural philosophy, I do not hesitate to express my decided opinion in favor of Peck's Ganot. The matter, style, and illustration eminently adapt the work to the popular wants.

From REV. SAMUEL McKINNEY, D.D., *Pres't Austin College, Huntsville, Texas.*

It gives me pleasure to commend it to teachers. I have taught some classes with it as our text, and must say, for simplicity of style and clearness of illustration, I have found nothing as yet published of equal value to the teacher and pupil.

From C. V. SPEAR, *Principal Maplewood Institute, Pittsfield, Mass.*

I am much pleased with its ample illustrations by plates, and its clearness and simplicity of statement. It covers the ground usually gone over by our higher classes, and contains many fresh illustrations from life or daily occurrences, and new applications of scientific principles to such.

From J. A. BANFIELD, *Superintendent Marshall Public Schools, Michigan.*

I have used Peck's Ganot since 1863, and with increasing pleasure and satisfaction each term. I consider it superior to any other work on physics in its adaptation to our high schools and academies. Its illustrations are superb—better than three times their number of pages of fine print.

From A. SCHUYLER, *Prof. of Mathematics in Baldwin University, Berea, Ohio.*

After a careful examination of Peck's Ganot's Natural Philosophy, and an actual test of its merits as a text-book, I can heartily recommend it as admirably adapted to meet the wants of the grade of students for which it is intended. Its diagrams and illustrations are *unrivaled*. We use it in the Baldwin University.

From D. C. VAN NORMAN, *Principal Van Norman Institute, New York.*

The Natural Philosophy of M. Ganot, edited by Prof. Peck, is, in my opinion, the best work of its kind, for the use intended, ever published in this country. Whether regarded in relation to the natural order of the topics, the precision and clearness of its definitions, or the fullness and beauty of its illustrations, it is certainly, I think, an advance.

☞ For many similar testimonials, see current numbers of the Illustrated Educational Bulletin.

CHEMISTRY.

Porter's First Book of Chemistry, ...$1 00
Porter's Principles of Chemistry, ... 2 00

The above are widely known as the productions of one of the most eminent scientific men of America. The extreme simplicity in the method of presenting the science, while exhaustively treated, has excited universal commendation. Apparatus adequate to the performance of every experiment mentioned, may be had of the publishers for a trifling sum. The effort to popularize the science is a great success. It is now within the reach of the poorest and least capable at once.

Darby's Text-Book of Chemistry, ... 1 75

Purely a Chemistry, divesting the subject of matters comparatively foreign to it (such as heat, light, electricity, etc.), but usually allowed to engross too much attention in ordinary school-books.

Gregory's Organic Chemistry, ... 2 50
Gregory's Inorganic Chemistry, ... 2 50

The science exhaustively treated. For colleges and medical students.

Steele's Fourteen Weeks' Course, ... 1 25

A successful effort to reduce the study to the limits of a *single term*, thereby making feasible its general introduction in institutions of every character   The author's felicity of style and success in making the science pre-eminently *interesting* are peculiarly noticeable features.

Chemical Apparatus, to accompany "Porter"   20 00
     do   do   to accompany "Steele"   25 00

BOTANY.

Thinker's First Lessons in Botany, ...   40

For children. The technical terms are largely dispensed with in favor of an easy and familiar style adapted to the smallest learner.

Wood's Object Lessons in Botany, ... 1 50
Wood's American Botanist and Florist, . . 2 50
Wood's New Class-Book of Botany, . . . 3 50

The standard text-books of the United States in this department. In style they are simple, popular, and lively; in arrangement, easy and natural; in description, graphic and strictly exact. The Tables for Analysis are reduced to a perfect system. More are annually sold than of all others combined.

Darby's Southern Botany, ... 2 00

Embracing general Structural and Physiological Botany, with vegetable products, and descriptions of Southern plants, and a complete Flora of the Southern States.

NATURAL SCIENCE—Continued

PHYSIOLOGY.

Jarvis' Elements of Physiology, ...$ 75

Jarvis' Physiology and Laws of Health, . 1 65

The only books extant which approach this subject with a proper view of the true object of teaching Physiology in schools, viz., that scholars may know how to take care of their own health. In bold contrast with the abstract *Anatomies*, which children learn as they would Greek or Latin (and forget as soon), to *discipline the mind*, are these text-books, using the *science* as a secondary consideration, and only so far as is necessary for the comprehension of the *laws of health*.

Hamilton's Vegetable & Animal Physiology, 1 25

The two branches of the science combined in one volume lead the student to a proper comprehension of the Analogies of Nature.

ASTRONOMY.

Steele's Fourteen Weeks' Course, ... 1 50

Reduced to a single term, and better adapted to school use than any work heretofore published. Not written for the information of scientific men, but for the inspiration of youth, the pages are not burdened with a multitude of figures which no memory could possibly retain. The whole subject is presented in a clear and concise form.

Willard's School Astronomy, ...1 00

By means of clear and attractive illustrations, addressing the eye in many cases by analogies, careful definitions of all necessary technical terms, a careful avoidance of verbiage and unimportant matter, particular attention to analysis, and a general adoption of the simplest methods, Mrs. Willard has made the best and most attractive *elementary* Astronomy extant.

McIntyre's Astronomy and the Globes, . . 1 50

A complete treatise for intermediate classes. Highly approved.

Bartlett's Spherical Astronomy, ...4 50

The West Point course, for advanced classes, with applications to the current wants of Navigation, Geography, and Chronology.

NATURAL HISTORY.

Carl's Child's Book of Natural History, . . 0 50

Illustrating the Animal, Vegetable, and Mineral Kingdoms, with application to the Arts. For beginners. Beautifully and copiously illustrated.

ZOOLOGY.

Chambers' Elements of Zoology, ...1 50

A complete and comprehensive system of Zoology, adapted for academic instruction, presenting a systematic view of the Animal Kingdom as a portion of external Nature.

Jarvis' Physiology and Laws of Health.

TESTIMONIALS.

From SAMUEL B. McLANE, *Superintendent Public Schools, Keokuk, Iowa,*

I am glad to see a really good text-book on this much neglected branch. This is clear, concise, accurate, and eminently adapted to the *class-room.*

From WILLIAM F. WYERS, *Principal of Academy, West Chester, Pennsylvania.*

A thorough examination has satisfied me of its superior claims as a text-book to the attention of teacher and taught. I shall introduce it at once.

From H. R. SANFORD, *Principal of East Genesee Conference Seminary, N. Y.*

"Jarvis' Physiology" is received, and fully met our expectations. We immediately adopted it.

From ISAAC T. GOODNOW, *State Superintendent of Kansas—published in connection with the "School Law."*

"Jarvis' Physiology," a common-sense, practical work, with just enough of anatomy to understand the physiological portions. The last six pages, on Man's Responsibility for his own health, are worth the price of the book.

From D. W. STEVENS, *Superintendent Public Schools, Fall River, Mass.*

I have examined Jarvis' "Physiology and Laws of Health," which you had the kindness to send to me a short time ago. In my judgment it is far the best work of the kind within my knowledge. It has been adopted as a text-book in our public schools.

From HENRY G. DENNY, *Chairman Book Committee, Boston, Mass.*

The very excellent "Physiology" of D. Jarvis I had introduced into our High School, where the study had been temporarily dropped, believing it to be by far the best work of the kind that had come under my observation; indeed, the reintroduction of the study was delayed for some months, because Dr. Jarvis' book could not be had, and we were unwilling to take any other.

From PROF. A. P. PEABODY, *D.D., LL.D., Harvard University.*

* * I have been in the habit of examining school-books with great care, and I hesitate not to say that, of all the text-books on Physiology which have been given to the public, Dr. Jarvis' deserves the first place on the score of accuracy, thoroughness, method, simplicity of statement, and constant reference to topics of practical interest and utility.

From JAMES N. TOWNSEND, *Superintendent Public Schools, Hudson, N. Y.*

Every human being is appointed to take charge of his own body; and of all books written upon this subject, I know of none which will so well prepare one to do this as "Jarvis' Physiology"—that is, in so small a compass of matter. It considers the pure, simple *laws of health* paramount to science; and though the work is thoroughly scientific, it is divested of all cumbrous technicalities, and presents the subject of physical life in a manner and style really charming. It is unquestionably the best text-book on physiology I have ever seen. It is giving great satisfaction in the schools of this city, where it has been adopted as the standard.

From L. J. SANFORD, *M.D., Prof. Anatomy and Physiology in Yale College*

Books on human physiology, designed for the use of schools, are more generally a failure perhaps than are school-books on most other subjects.

The great want in this department is met, we think, in the well-written treatise of Dr. Jarvis, entitled "Physiology and Laws of Health." * * The work is not too detailed nor too expansive in any department, and is clear and concise in all. It is not burdened with an excess of anatomical description, nor rendered discursive by many zoological references. Anatomical statements are made to the extent of qualifying the student to attend, understandingly, to an exposition of those functional processes which, collectively, make up health; thus the laws of health are enunciated, and many suggestions are given which, if heeded, will tend to its preservation.

☞ For further testimony of similar character, see current numbers of the Illustrated Educational Bulletin.

NATURAL SCIENCE.

"FOURTEEN WEEKS" IN EACH BRANCH.
By J. DORMAN STEELE, A. M.

Steele's 14 Weeks Course in Chemistry . $1 25
Steele's 14 Weeks Course in Astronomy . 1 50
Steele's 14 Weeks Course in Philosophy . 1 50
Steele's 14 Weeks Course in Geology.

The unparalleled success of the first volume, "14 *Weeks in Chemistry*," encouraged the publishers to project a complete course upon a similar plan, and designed to make the Natural Sciences *popular*.

Our Text-Books in these studies are, as a general thing, dull and uninteresting. They contain from 400 to 600 pages of dry facts and unconnected details. They abound in that which the student cannot learn, much less remember. The pupil commences the study, is confused by the fine print and coarse print, and neither knowing exactly what to learn nor what to hasten over, is crowded through the single term generally assigned to each branch, and frequently comes to the close without a definite and exact idea of a single scientific principle.

Steele's Fourteen Weeks Courses contain only that which every well-informed person should know, while all that which concerns only the professional scientist is omitted. The language is clear, simple, and interesting, and the illustrations bring the subject within the range of home life and daily experience. They give such of the general principles and the prominent facts as a pupil can make familiar as household words within a single term. The type is large and open; there is no fine print to annoy; the cuts are copies of genuine experiments or natural phenomena, and are of fine execution.

In fine, by a system of condensation peculiarly his own, the author reduces each branch to the limits of a single term of study, while sacrificing nothing that is essential, and nothing that is usually retained from the study of the larger manuals in common use. Thus the student has rare opportunity to *economize his time*, or rather to employ that which he has to the best advantage.

A notable feature is the author's charming "style," fortified by an enthusiasm over his subject in which the student will not fail to partake. Believing that Natural Science is full of fascination, he has moulded it into a form that attracts the attention and kindles the enthusiasm of the pupil.

The recent editions contain the author's "Practical Questions" on a plan never before attempted in scientific text-books. These are questions as to the nature and cause of common phenomena, and are not directly answered in the text, the design being to test and promote an intelligent use of the student's knowledge of the foregoing principles.

Steele's General Key to his Works. *1 50

This work is mainly composed of Answers to the Practical Questions and Solutions of the Problems in the author's celebrated "Fourteen Weeks Courses" in the several sciences, with many hints to teachers, minor Tables, &c. Should be on every teacher's desk.

Steele's 14 Weeks in each Science.

TESTIMONIALS.

From L. A. Bikle, President N. C. College.

I have not been disappointed. Shall take pleasure in introducing this series.

From J. F. Cox, Prest. Southern Female College, Ga.

I am much pleased with these books, and expect to introduce them.

From J. R. Branham, Prin. Brownsville Female College, Tenn.

They are capital little books, and are now in use in our institution.

From W. H. Goodale, Professor Readville Seminary, La.

We are using your 14 Weeks Course, and are much pleased with them.

From W. A. Boles, Supt. Shelbyville Graded School, Ind.

They are as entertaining as a story book, and much more improving to the mind.

From S. A. Snow, Principal of High School, Uxbridge, Mass.

Steele's 14 Weeks Courses in the Sciences are a perfect success.

From John W. Doughty, Newburg Free Academy, N. Y.

I was prepared to find Prof. Steele's Course both attractive and instructive. My **highest** expectations have been fully realized.

From J. S. Blackwell, Prest. Ghent College, Ky.

Prof. Steele's unexampled success in providing for the wants of academic classes, has led me to look forward with high anticipations to his forthcoming issue.

From J. F. Cook, Prest. La Grange College, Mo.

I am pleased with the neatness of these books and the delightful diction. I have been teaching for years, and have never seen a lovelier little volume than the Astronomy.

From M. W. Smith, Prin. of High School, Morrison, Ill.

They seem to me to be admirably adapted to the wants of a public school, containing, as they do, a sufficiently comprehensive arrangement of elementary principles to excite a healthy thirst for a more thorough knowledge of those sciences.

From J. D. Bartley, Prin. of High School, Concord, N H.

They are just such books as I have looked for, viz., those of interesting style, not cumbersome and filled up with things to be omitted by the pupil, and yet sufficiently full of facts for the purpose of most scholars in these sciences in our high schools; there is nothing but what a pupil of average ability can thoroughly master.

From Alonzo Norton Lewis, Principal of Parker Academy, Conn.

I consider Steele's Fourteen Weeks Courses in Philosophy, Chemistry, &c., the *best* school-books that have been issued in this country.

As an introduction to the various branches of which they treat, and especially for that numerous class of pupils who have not the time for a more extended course, I consider them *invaluable*.

From Edward Brooks, Prin. State Normal School, Millersville, Pa.

At the meeting of Normal School Principals, I presented the following resolution, which was unanimously adopted: "*Resolved*, That Steele's 14 Weeks Courses in Natural Philosophy and Astronomy, or an amount equivalent to what is contained in them, be adopted for use in the State Normal Schools of Pennsylvania." The works themselves will be adopted by at least three of the schools, and, I presume, by them all.

MODERN LANGUAGE.

French and English Primer, ...$ 10
German and English Primer, ... 10
Spanish and English Primer, ... 10

The names of common objects properly illustrated and arranged in easy lessons.

Ledru's French Fables, ... 75
Ledru's French Grammar, ... 1 00
Ledru's French Reader, ... 1 00

The author's long experience has enabled him to present the most thoroughly practical text-books extant, in this branch. The system of pronunciation (by phonetic illustration) is original with this author, and will commend itself to all American teachers, as it enables their pupils to secure an absolutely correct pronunciation without the assistance of a native master. This feature is peculiarly valuable also to "self-taught" students. The directions for ascertaining the gender of French nouns—also a great stumbling-block—are peculiar to this work, and will be found remarkably competent to the end proposed. The criticism of teachers and the test of the school-room is invited to this excellent series, with confidence.

Worman's French Echo, ... 1 25

To teach conversational French by actual practice, on an entirely new plan, which recognizes the importance of the student learning *to think* in the language which he speaks. It furnishes an extensive vocabulary of words and expressions in common use, and suffices to free the learner from the embarrassments which the peculiarities of his own tongue are likely to be to him, and to make him thoroughly familiar with the use of proper idioms.

Worman's German Echo, ... 1 25

On the same plan. See Worman's German Series, page 29.

Pujol's Complete French Class-Book, . . . 2 25

Offers, in one volume, methodically arranged, a complete French course —usually embraced in series of from five to twelve books, including the bulky and expensive Lexicon. Here are Grammar, Conversation, and choice Literature—selected from the best French authors. Each branch is thoroughly handled; and the student, having diligently completed the course as prescribed, may consider himself, without further application, *au fait* in the most polite and elegant language of modern times.

Maurice-Poitevin's Grammaire Francaise, . 1 00

American schools are at last supplied with an American edition of this famous text-book. Many of our best institutions have for years been procuring it from abroad rather than forego the advantages it offers. The policy of putting students who have acquired some proficiency from the ordinary text-books, into a Grammar written in the vernacular, can not be too highly commended. It affords an opportunity for finish and review at once; while embodying abundant practice of its own rules.

Willard's Historia de los Estados Unidos, . 2 00

The History of the United States, translated by Professors TOLON and DE TORNOS, will be found a valuable, instructive, and entertaining reading-book for Spanish classes.

Pujol's Complete French Class-Book.

TESTIMONIALS.

From Prof. Elias Peissner, Union College.

I take great pleasure in recommending Pujol and Van Norman's French Class-Book, as there is no French grammar or class-book which can be compared with it in completeness, system, clearness, and general utility.

From Edward North, President of Hamilton College.

I have carefully examined Pujol and Van Norman's French Class-Book, and am satisfied of its superiority, for college purposes, over any other heretofore used. We shall not fail to use it with our next class in French.

From A. Curtis, Pres't of Cincinnati Literary and Scientific Institute.

I am confident that it may be made an instrument in conveying to the student, in from six months to a year, the art of speaking and writing the French with almost native fluency and propriety.

From Hiram Orcutt, A. M., Prin. Glenwood and Tilden Ladies' Seminaries.

I have used Pujol's French Grammar in my two seminaries, exclusively, for more than a year, and have no hesitation in saying that I regard it the best text-book in this department extant. And my opinion is confirmed by the testimony of Prof. F De Launay and Mademoiselle Marindin. They assure me that the book is eminently accurate and practical, as tested in the school-room.

From Prof. Theo. F. De Fumat, Hebrew Educational Institute, Memphis, Tenn.

M. Pujol's French Grammar is one of the best and most practical works. The French language is chosen and elegant in style—modern and easy. It is far superior to the other French class-books in this country. The selection of the conversational part is very good, and will interest pupils; and being all completed in only one volume, it is especially desirable to have it introduced in our schools.

From Prof. James H. Worman, Bordentown Female College, N. J.

The work is upon the same plan as the text-books for the study of French and English published in Berlin, for the study of those who have not the aid of a teacher, and these books are considered, by the first authorities, the best books. In most of our institutions, Americans teach the modern languages, and heretofore the trouble has been to give them a text-book that would dispose of the difficulties of the French pronunciation. This difficulty is successfully removed by P and Van N., and I have every reason to believe it will soon make its way into most of our best schools.

From Prof. Charles S. Dod, Ann Smith Academy, Lexington, Va.

I cannot do better than to recommend "Pujol and Van Norman." For comprehensive and systematic arrangement, progressive and thorough development of all grammatical principles and idioms, with a due admixture of theoretical knowledge and practical exercise, I regard it as superior to any (other) book of the kind.

From A. A. Forster, Prin. Pinehurst School, Toronto, C. W.

I have great satisfaction in bearing testimony to M. Pujol's System of French Instruction, as given in his complete class-book. For clearness and comprehensiveness, adapted for all classes of pupils, I have found it superior to any other work of the kind, and have now used it for some years in my establishment with great success.

From Prof. Otto Fedder, Maplewood Institute, Pittsfield, Mass.

The conversational exercises will prove an immense saving of the hardest kind of labor to teachers. There is scarcely any thing more trying in the way of teaching language, than to rack your brain for short and easily intelligible bits of conversation, and to repeat them time and again with no better result than extorting at long intervals a doubting " oui," or a hesitating " non, monsieur."

☞ For further testimony of a similar character, see special circular, and current numbers of the Educational Bulletin.

28

GERMAN.

A COMPLETE COURSE IN THE GERMAN.
By JAMES H. WORMAN, A. M.

Worman's Elementary German Grammar . $1 50
Worman's Complete German Grammar . 2 00

These volumes are designed for intermediate and advanced classes respectively. The bitterness with which they have been attacked, and their extraordinary success in the face of an unprincipled opposition, are facts which have stamped them as possessing unparalleled merit.

Though following the same general method with "Otto" (that of 'Caspey'), our author differs essentially in its application. He is more practical, more systematic, more accurate, and besides introduces a number of invaluable features which have never before been combined in a German grammar.

Among other things, it may be claimed for Prof. Worman that he has been *the first* to introduce in an American text-book for learning German, a system of analogy and comparison with other languages. Our best teachers are also enthusiastic about his methods of inculcating the art of speaking, of understanding the spoken language, of correct pronunciation; the sensible and convenient original classification of nouns (in four declensions), and of irregular verbs, also deserves much praise. We also note the use of heavy type to indicate etymological changes in the paradigms, and, in the exercises, the parts which specially illustrate preceding rules.

Worman's German Reader $1 75

The finest compilation of classical and standard German Literature ever offered to American students. It embraces, progressively arranged, selections from the masterpieces of Goethe, Schiller, Korner, Seume, Uhland, Freiligrath, Heine, Schlegel, Holty, Lenau, Wieland, Herder, Lessing, Kant, Fichte, Schelling, Winkelmann, Humboldt, Ranke, Raumer, Menzel, Gervinus, &c., and contains complete Goethe's "Iphigenie," Schiller's "Jungfrau;" also, for instruction in modern conversational German, Benedix's "Eigensinn."

There are besides, Biographical Sketches of each author contributing, Notes, explanatory and philological (after the text), Grammatical References to all leading grammars, as well as the editor's own, and an adequate Vocabulary.

Worman's German Echo $1 25

Consists of exercises in colloquial style entirely in the German, with an adequate vocabulary, not only of words but of idioms. The object of the system developed in this work (and its companion volume in the French) is to break up the laborious and tedious habit of *translating the thoughts*, which is the student's most effectual bar to fluent conversation, and to lead him to *think in the language in which he speaks*. As the exercises illustrate scenes in actual life, a considerable knowledge of the manners and customs of the German people is also acquired from the use of this manual.

Worman's German Grammars.

TESTIMONIALS.

From Prof. R. W. JONES, *Petersburg Female College, Va.*

From what I have seen of the work it is almost certain *I shall introduce it* into this institution.

From Prof. G. CAMPBELL, *University of Minnesota.*

A valuable addition to our school-books, and will find many friends, and do great good.

From Prof. O. H P. CORPREW, *Mary Military Inst., Md.*

I am better pleased with them than any I have ever taught. I have already ordered through our booksellers.

From Prof. R. S. KENDALL, *Vernon Academy, Conn.*

I at once put the Elementary Grammar into the hands of a class of beginners, and have used it *with great satisfaction.*

From Prof. D. E. HOLMES, *Berlin Academy, Wis.*

Worman's German works are *superior.* I shall use them hereafter in my German classes.

From Prof. MAGNUS BUCHNOLTZ, *Hiram College, Ohio.*

I have examined the Complete Grammar, and find it *excellent.* You may rely that it will be used here.

From Prin. THOS. W. TOBEY, *Paducah Female Seminary, Ky.*

The Complete German Grammar is worthy of an extensive circulation. It is *admirably adapted* to the class-room. I shall use it.

From Prof. ALEX. ROSENSPITZ, *Houston Academy, Texas.*

Bearer will take and pay for 3 dozen copies. Mr. Worman deserves the approbation and esteem of the teacher and the thanks of the student.

From Prof. G. MALMENE, *Augusta Seminary, Maine.*

The Complete Grammar cannot fail to *give great satisfaction* by the simplicity of its arrangement, and by its completeness.

From Prin. OVAL PIRKEY, *Christian University, Mo.*

Just such a series as is positively necessary. I do hope the author will succeed as well in the French, &c., as he has in the German.

From Prof. S. D. HILLMAN, *Dickinson College, Pa.*

The class have lately commenced, and my examination thus far warrants me in saying that I regard it as *the best* grammar for instruction in the German.

From Prin. SILAS LIVERMORE, *Bloomfield Seminary, Mo.*

I have found a classically and scientifically educated Prussian gentleman whom I propose to make German instructer. I have shown him both your German grammars. He has expressed *his approbation* of them generally.

From Prof. Z. TEST, *Howland School for Young Ladies, N. Y.*

I shall introduce the books. From a cursory examination I have no hesitation in pronouncing the Complete Grammar *a decided improvement* on the text-books at present in use in this country.

From Prof. LEWIS KISTLER, *Northwestern University, Ill.*

Having looked through the Complete Grammar with some care I must say that you have produced *a good book;* you may be rewarded with this gratification—that your grammar promotes the facility of learning the German language, and of becoming acquainted with its rich literature.

From Pres. J. P. ROVS, *Stockwell Collegiate Inst., Ind.*

I supplied a class with the Elementary Grammar, and it gives *complete satisfaction.* The conversational and reading exercises are well calculated to illustrate the principles, and lead the student on an easy yet thorough course. I think the Complete Grammar equally attractive.

THE CLASSICS.

LATIN.

Silber's Latin Course, ... $1 25

The book contains an Epitome of Latin Grammar, followed by Reading Exercises, with explanatory Notes and copious References to the leading Latin Grammars, and also to the Epitome which precedes the work. Then follow a Latin-English Vocabulary and Exercises in Latin Prose Composition, being thus complete in itself, and a very suitable work to put in the hands of one about to study the language.

Searing's Virgil's Æneid, ... 2 25

It contains only the first six books of the Æneid. 2. A very carefully constructed Dictionary. 3. Sufficiently copious Notes. 4. Grammatical references to four leading Grammars. 5. Numerous Illustrations of the highest order. 6. A superb Map of the Mediterranean and adjacent countries. 7. Dr. S. H. Taylor's "Questions on the Æneid." 8. A Metrical Index, and an Essay on the Poetical Style. 9. A photographic *fac simile* of an early Latin M.S. 10. The text according to Jahn, but paragraphed according to Ladewig. 11. Superior mechanical execution.

Hanson's Latin Prose Book, ... 3 00
Hanson's Latin Poetry, ... 3 00

Andrews & Stoddard's Latin Grammar, *1 50
Andrews' Questions on the Grammar, . *0 15
Andrews' Latin Exercises, ... *1 25
Andrews' Viri Romæ, ... *1 25
Andrews' Sallust's Jugurthine War, &c. *1 50
Andrews' Eclogues & Georgics of Virgil, *1 50
Andrews' Cæsar's Commentaries, ... *1 50
Andrews' Ovid's Metamorphoses, . . . *1 25

GREEK.

Crosby's Greek Grammar, ... 2 00
Crosby's Xenophon's Anabasis, ... 1 25

MYTHOLOGY.

Dwight's Grecian and Roman Mythology.

School edition, $1 25;   University edition,   *3 00

A knowledge of the fables of antiquity, thus presented in a systematic form, is as indispensable to the student of general literature as to him who would peruse intelligently the classical authors. The mythological allusions so frequent in literature are readily understood with such a Key as this.

ELOCUTION.

Watson's Practical Elocution $0 25
A brief, clear, and most satisfactory treatise—same as in "**Independent Fifth Reader.**" The subject fully illustrated by diagrams.

Zachos' Analytic Elocution 1 50
All departments of elocution—such as the analysis of the voice and the sentence, phonology, rhythm, expression, gesture, &c.—are here arranged for instruction in classes, illustrated by copious examples.

Sherwood's Self Culture 1 00
Self-culture in reading, speaking, and conversation—a very valuable treatise to those who would perfect themselves in these accomplishments.

SPEAKERS.

Northend's Little Orator *60
Contains simple and attractive pieces in prose and poetry, adapted to the capacity of children under twelve years of age.

Northend's National Orator *1 25
About one hundred and seventy choice pieces happily arranged. The design of the author in making the selection has been to cultivate *versatility of expression.*

Northend's Entertaining Dialogues *1 25
Extracts eminently adapted to cultivate the dramatic faculties, as well as entertain an audience.

Swett's Common School Speaker *1 25
Selections from recent literature.

Raymond's Patriotic Speaker *2 00
A superb compilation of modern eloquence and poetry, with original dramatic exercises. Nearly every eminent *living* orator is represented, without distinction of place or party.

COMPOSITION, &c.

Brookfield's First Book in Composition . 50
Making the cultivation of this important art feasible for the smallest child. By a new method, to induce and stimulate thought.

Boyd's Composition and Rhetoric 1 50
This work furnishes all the aid that is needful or can be desired in the various departments and styles of composition, both in prose and verse.

Day's Art of Rhetoric 1 25
Noted for exactness of definition, clear limitation, and philosophical development of subject; the large share of attention given to Invention, as a branch of Rhetoric, and the unequalled analysis of style

LITERATURE.

Boyd's Milton's Paradise Lost *1 25

Boyd's Young's Night Thoughts *1 25

Boyd's Cowper's Task, Table Talk, &c. .*1 25

Boyd's Thomson's Seasons *1 25

Boyd's Pollok's Course of Time *1 25

Boyd's Lord Bacon's Essays *1 60

> This series of annotated editions of great English writers, in prose and poetry, is designed for critical reading and parsing in schools. Prof. J. R. Boyd proves himself an editor of high capacity, and the works themselves need no encomium. As auxiliary to the study of Belles Lettres, etc., these works have no equal.

Pope's Essay on Man *20

Pope's Homer's Iliad *80

> The metrical translation of the great poet of antiquity, and the matchless "Essay on the Nature and State of Man," by ALEXANDER POPE, afford superior exercise in literature and parsing.

AESTHETICS.

Huntington's Manual of the Fine Arts . .*1 75

> A view of the rise and progress of Art in different countries, a brief account of the most eminent masters of Art, and an analysis of the principles of Art. It is complete in itself, or may precede to advantage the critical work of Lord Kames.

Boyd's Kames' Elements of Criticism . .*1 75

> The best edition of this standard work; without the study of which none may be considered proficient in the science of the Perceptions. No other study can be pursued with so marked an effect upon the taste and refinement of the pupil.

POLITICAL ECONOMY.

Champlin's Lessons on Political Economy 1 25

> An improvement on previous treatises, being shorter, yet containing every thing essential, with a view of recent questions in finance, etc., which is not elsewhere found.

MENTAL PHILOSOPHY.

Mahan's Intellectual Philosophy $1 75

The subject exhaustively considered. The author has evinced learning, candor, and independent thinking.

Mahan's Science of Logic 2 00

A profound analysis of the laws of thought. The system possesses the merit of being intelligible and self consistent. In addition to the author's carefully elaborated views, it embraces results attained by the ablest minds of Great Britain, Germany, and France, in this department.

Boyd's Elements of Logic 1 25

A systematic and philosophic condensation of the subject, fortified with additions from Watts, Abercrombie, Whately, &c.

Watts on the Mind 50

The Improvement of the Mind, by Isaac Watts, is designed as a guide for the attainment of useful knowledge. As a text-book it is unparalleled; and the discipline it affords cannot be too highly esteemed by the educator.

MORALS.

Alden's Text-Book of Ethics 60

For young pupils. To aid in systematizing the ethical teachings of the Bible, and point out the coincidences between the instructions of the sacred volume and the sound conclusions of reason.

Willard's Morals for the Young *75

Lessons in conversational style to inculcate the elements of moral philosophy. The study is made attractive by narratives and engravings.

GOVERNMENT.

Howe's Young Citizen's Catechism 75

Explaining the duties of District, Town, City, County, State, and United States Officers, with rules for parliamentary and commercial business—that which every future " sovereign" ought to know, and so few are taught.

Young's Lessons in Civil Government . . 1 25

A comprehensive view of Government, and abstract of the laws showing the rights, duties, and responsibilities of citizens.

Mansfield's Political Manual 1 25

This is a complete view of the theory and practice of the General and State Governments of the United States, designed as a text-book. The author is an esteemed and able professor of constitutional law, widely known for his sagacious utterances in matters of statecraft through the public press. Recent events teach with emphasis the vital necessity that the rising generation should comprehend the noble polity of the American government, that they may act intelligently when endowed with a voice in it.

TEACHERS' AIDS.

Brooks' School Manual of Devotion . . . 75

This volume contains daily devotional exercises, consisting of a hymn, selections of scripture for alternate reading by teacher and pupils, and a prayer. Its value for opening and closing school is apparent.

Cleaveland's School Harmonist *70

Contains appropriate *tunes* for each hymn in the "Manual of Devotion" described above.

The Boy Soldier 75

Complete infantry tactics for schools, with illustrations, for the use of those who would introduce this pleasing relaxation from the confining duties of the desk.

Welch's Object Lessons 1 00

Invaluable for teachers of primary schools. Contains the best explanation of the Pestalozzian system. By its aid the proficiency of pupils and the general interest of the school may be increased one hundred per cent.

Tracy's School Record *75

To record attendance, deportment, and scholarship; containing also many useful tables and suggestions to teachers, that are worth of themselves the price of the book.

Tracy's Pocket Record *65

A portable edition of the School Record, without the tables, &c.

Brooks' Teacher's Register *1 00

Presents at one view a record of attendance, recitations, and deportment for the whole term.

Carter's Record and Roll-Book *2 50

For large graded schools.

National School Diary, per dozen *1 00

A little book of blank forms for weekly report of the standing of each scholar, from teacher to parent. A great convenience.

THE
TEACHER'S LIBRARY.

The Student; or, Fireside Friend — Phelps, $*1.50
The Educator; or, Hours with my
    Pupils, ... do., *1 50
The Discipline of Life; or, Ida
    Norman, ... do., *1 75

> The authoress of these works is one of the most distinguished writers on education; and they can not fail to prove a valuable addition to the School and Teachers' Libraries, being in a high degree both interesting and instructive.

Hecker's Scientific Basis of Education, . .*2 50

> Adaptation of study and classification by temperaments.

Object Lessons—Welch *1 00

> This is a complete exposition of the popular modern system of "object-teaching," for teachers of primary classes.

Theory and Practice of Teaching—Page .*1 50

> This volume has, without doubt, been read by two hundred thousand teachers, and its popularity remains undiminished—large editions being exhausted yearly. It was the pioneer, as it is now the patriarch of professional works for teachers.

The Graded School—Wells *1 25

> The proper way to organize graded schools is here illustrated. The author has availed himself of the best elements of the several systems prevalent in Boston, New York, Philadelphia, Cincinnati, St. Louis, and other cities.

The Normal—Holbrook *1 75

> Carries a working school on its visit to teachers, showing the most approved methods of teaching all the common branches, including the technicalities, explanations, demonstrations, and definitions introductory and peculiar to each branch.

The Teachers' Institute—Fowle *1 25

> This is a volume of suggestions inspired by the author's experience at institutes, in the instruction of young teachers. A thousand points of interest to this class are most satisfactorily dealt with.

The Teacher and the Parent—Northend . $*1 50

A treatise upon common-school education, designed to lead teachers to view their calling in its true light, and to stimulate them to fidelity.

The Teachers' Assistant—Northend . . . *1 50

A natural continuation of the author's previous work, more directly calculated for daily use in the administration of school discipline and instruction.

School Government—Jewell *1 50

Full of advanced ideas on the subject which its title indicates. The criticisms upon current theories of punishment and schemes of administration have excited general attention and comment.

Grammatical Diagrams—Jewell *1 00

The diagram system of teaching grammar explained, defended, and improved. The curious in literature, the searcher for truth, those interested in new inventions, as well as the disciples of Prof. Clark, who would see their favorite theory fairly treated, all want this book. There are many who would like to be made familiar with this system before risking its use in a class. The opportunity is here afforded.

The Complete Examiner—Stone *1 25

Consists of a series of questions on every English branch of school and academic instruction, with reference to a given page or article of leading text-books where the answer may be found in full. Prepared to aid teachers in securing certificates, pupils in preparing for promotion, and teachers in selecting review questions.

School Amusements—Root *1 50

To assist teachers in making the school interesting, with hints upon the management of the school-room. Rules for military and gymnastic exercises are included. Illustrated by diagrams.

Institute Lectures on Mental and Moral Culture—Bates *1 50

These lectures, originally delivered before institutes, are based upon various topics of interest to the teacher. The volume is calculated to prepare the will, awaken the inquiry, and stimulate the thought of the zealous teacher.

Method of Teachers' Institutes—Bates . . . * 75

Sets forth the best method of conducting institutes, with a detailed account of the object, organization, plan of instruction, and true theory of education on which such instruction should be based.

History and Progress of Education . . . *1 50

The systems of education prevailing in all nations and ages, the gradual advance to the present time, and the bearing of the past upon the present in this regard, are worthy of the careful investigation of all concerned in the cause.

American Education—Mansfield $1 50

A treatise on the principles and elements of education, as practiced in this country, with ideas towards distinctive republican and Christian education.

American Institutions—De Tocqueville . .*1 50

A valuable index to the genius of our Government.

Universal Education—Mayhew *1 75

The subject is approached with the clear, keen perception of one who has observed its necessity, and realized its feasibility and expediency alike. The redeeming and elevating power of improved common schools constitutes the inspiration of the volume.

Higher Christian Education—Dwight . . .*1 50

A treatise on the principles and spirit, the modes, directions, and results of all true teaching; showing that right education should appeal to every element of enthusiasm in the teacher's nature.

Modern Philology—Dwight *1 75

Important to the grammarian, and indispensable to the teacher of language, ancient or modern, who would afford his pupils the advantage of the analogy and association to be derived from an intelligent comparison of all languages and their history.

Lectures on Natural History—Chadbourne * 75

Affording many themes for oral instruction in this interesting science—especially in schools where it is not pursued as a class exercise.

Outlines of Mathematical Science—Davies *1 00

A manual suggesting the best methods of presenting mathematical instruction on the part of the teacher, with that comprehensive view of the whole which is necessary to the intelligent treatment of a part, in science.

Logic & Utility of Mathematics—Davies . .*1 50

An elaborate and lucid exposition of the principles which lie at the foundation of pure mathematics, with a highly ingenious application of their results to the development of the essential idea of the different branches of the science.

Mathematical Dictionary—Davies & Peck .*3 50

This cyclopædia of mathematical science defines with completeness, precision, and accuracy, every technical term, thus constituting a popular treatise on each branch, and a general view of the whole subject.

School Architecture—Barnard *2 25

Attention is here called to the vital connection between a good school-house and a good school, with plans and specifications for securing the former in the most economical and satisfactory manner.

THE SCHOOL LIBRARY.

The two elements of instruction and entertainment were never more happily combined than in this collection of standard books. Children and adults alike will here find ample food for the mind, of the sort that is easily *digested*, while not degenerating to the level of modern romance.

LIBRARY OF LITERATURE.

Milton's Paradise Lost   Boyd's Illustrated Ed. $1 60

Young's Night Thoughts . . do. . . 1 60

Cowper's Task, Table Talk, &c. . do. . . 1 60

Thomson's Seasons . . . . . do. . . 1 60

Pollok's Course of Time . . . do. . . 1 60

These great moral poems are known wherever the English language is read, and are regarded as models of the best and purest literature. The books are beautifully illustrated, and notes explain all doubtful meanings, and furnish other matter of interest to the general reader.

Lord Bacon's Essays, (Boyd's Edition.) . . . 1 60

Another grand English classic, affording the highest example of purity in language and style.

The Iliad of Homer. Translated by POPE. . . 80

Those who are unable to read this greatest of ancient writers in the original, should not fail to avail themselves of this metrical version by an eminent scholar and poet.

The Poets of Connecticut—Everest 1 75

With the biographical sketches, this volume forms a complete history of the poetical literature of the State.

The Son of a Genius—Hofland 75

A juvenile classic which never wears out, and finds many interested readers in every generation of youth.

Lady Willoughby 1 00

The diary of a wife and mother. An historical romance of the seventeenth century. At once beautiful and pathetic, entertaining and instructive.

The Rhyming Dictionary—Walker 1 25

A serviceable manual to composers of rhythmical matter, being a complete index of allowable rhymes.

LIBRARY OF REFERENCE.

Home Cyclopædia of Chronology · · · · .$2 25

An index to the sources of knowledge—a dictionary of dates.

Home Cyclopædia of Geography · · · · · 2 25

A complete gazetteer of the world.

Home Cyclopædia of Useful Arts· · · · · 2 25

Covering the principles and practice of modern scientific enterprise, with a record of important inventions in agriculture, architecture, domestic economy, engineering, machinery, manufactures, mining, photogenic and telegraphic art, &c., &c.

Home Cyclopædia of Literature & Fine Arts 2 25

A complete index to all terms employed in belles lettres, philosophy, theology, law, mythology, painting, music, sculpture, architecture, and all kindred arts.

LIBRARY OF TRAVEL.

Ship and Shore—Colton · · · · · · · · 1 50

In Madeira, Lisbon, and the Mediterranean Ocean. Illustrated.

Land and Lee—Colton · · · · · · · · 1 50

In the Bosphorus and Ægean. Illustrated.

Sea and Sailor—Colton · · · · · · · · 1 50

Notes on France and Italy. Illustrated.

Deck and Port—Colton · · · · · · · · 1 50

A cruise to California. Illustrated.

Three Years in California—Colton · · · · 1 50

During the gold fever. Illustrated.

These racy descriptions of travel are regarded as models in this department of literature. They are read by old and young with vast interest and profit.

A Visit to Europe—Silliman, 2 vols. · · · · 3 00

A very spicy book of foreign travel. It brings every opportunity of the tourist to the feet of the reader.

TRAVEL—Continued.

Life in the Sandwich Islands—Cheever . .$1 50

The "heart of the Pacific, as it was and is," shows most vividly the contrast between the depth of degradation and barbarism, and the light and liberty of civilization, so rapidly realized in these islands under the humanizing influence of the Christian religion. Illustrated.

Peruvian Antiquities—Von Tschudi 1 50

Travels in Peru—Von Tschudi 1 50

The first of these volumes affords whatever information has been attained by travelers and men of science concerning the extinct people who once inhabited Peru, and who have left behind them many relics of a wonderful civilization. The "Travels" furnish valuable information concerning the country and its inhabitants as they now are. Illustrated.

Ancient Monasteries of the East—Curzon . 1 50

The exploration of these ancient seats of learning has thrown much light upon the researches of the historian, the philologist, and the theologian, as well as the general student of antiquity. Illustrated.

Discoveries in Babylon & Nineveh—Layard 1 75

Valuable alike for the information imparted with regard to these most interesting ruins, and the pleasant adventures and observations of the author in regions that to most men seem like Fairyland. Illustrated.

A Run Through Europe—Benedict, ... 2 00

A work replete with instruction and interest.

St. Petersburgh—Jermann 1 00

Americans are less familiar with the history and social customs of the Russian people than those of any other modern civilized nation. Opportunities such as this book affords are not, therefore, to be neglected.

The Polar Regions—Osborn 1 25

A thrilling and intensely interesting narrative of one of the famous expeditions in search of Sir John Franklin—unsuccessful in its main object, but adding many facts to the repertoire of science.

Thirteen Months in the Confederate Army 75

The author, a northern man conscripted into the Confederate service, and rising from the ranks by soldierly conduct to positions of responsibility, had remarkable opportunities for the acquisition of facts respecting the conduct of the Southern armies, and the policy and deeds of their leaders. He participated in many engagements, and his book is one of the most exciting narratives of adventure ever published. Mr. Stevenson takes no ground as a partizan, but views the whole subject as with the eye of a neutral—only interested in subserving the ends of history by the contribution of impartial facts. Illustrated.

LIBRARY OF HISTORY.

History of Europe—Alison $2 50

A reliable and standard work. which covers with clear, connected, and complete narrative. the eventful occurrences transpiring from A. D. 1789 to 1815, being mainly a history of the career of Napoleon Bonaparte.

History of England—Berard 1 75

Combining a history of the social life of the English people with that of the civil and military transactions of the realm.

History of Rome—Ricord 1 60

Possesses all the charm of an attractive romance. The fables with which this history abounds are introduced in such away as not to deceive the inexperienced reader, while adding vastly to the interest of the work and affording a pleasing index to the genius of the Roman people. Illustrated.

The Republic of America—Willard 2 25

Universal History in Perspective—Willard 2 25

From these two comparatively brief treatises the intelligent mind may obtain a comprehensive knowledge of the history of the world in both hemispheres. Mrs. Willard's reputation as an historian is wide as the land. Illustrated.

Ecclesiastical History—Marsh 2 00

A history of the Church in all ages. with a comprehensive review of all forms of religion from the creation of the world. No other source affords, in the same compass, the information here conveyed.

History of the Ancient Hebrews—Mills . . 1 75

The record of "God's people" from the call of Abraham to the destruction of Jerusalem; gathered from sources sacred and profane.

The Mexican War—Mansfield 1 50

A history of its origin, and a detailed account of its victories; with official dispatches, the treaty of peace, and valuable tables. Illustrated.

Early History of Michigan—Sheldon . . . 1 75

A work of value and deep interest to the people of the West. Compiled under the supervision of Hon. Lewis Cass. Embellished with portraits.

LIBRARY OF BIOGRAPHY.

Life of Dr. Sam. Johnson—Boswell . .$2 25

This work has been before the public for seventy years, with increasing approbation.   Boswell is known as " the prince of biographers."

Henry Clay's Life and Speeches— Mallory
2 vols. 4 50

This great American statesman commands the admiration, and his character and deeds solicit the study of every patriot.

Life & Services of General Scott—Mansfield 1 75

The hero of the Mexican war, who was for many years the most prominent figure in American military circles, should not be forgotten in the whirl of more recent events than those by which he signalized himself. Illustrated.

Garibaldi's Autobiography 1 50

The Italian patriot's record of his own life, translated and edited by his friend and admirer. A thrilling narrative of a romantic career. With portrait.

Lives of the Signers—Dwight 1 50

The memory of the noble men who declared our country free at the peril of their own "lives, fortunes, and sacred honor," should be embalmed in every American's heart.

Life of Sir Joshua Reynolds—Cunningham 1 50

A candid, truthful, and appreciative memoir of the great painter, with a compilation of his discourses. The volume is a text-book for artists, as well as those who would acquire the rudiments of art. With a portrait.

Prison Life 75

Interesting biographies of celebrated prisoners and martyrs, designed especially for the instruction and cultivation of youth.

LIBRARY OF NATURAL SCIENCE.

The Treasury of Knowledge $1 25

A cyclopædia of ten thousand common things, embracing the widest range of subject-matter. Illustrated.

Ganot's Popular Physics 1 75

The elements of natural philosophy for both student and the general reader. The original work is celebrated for the magnificent character of its illustrations, all of which are literally reproduced here.

Principles of Chemistry—Porter 2 00

A work which commends itself to the amateur in science by its extreme simplicity, and careful avoidance of unnecessary detail. Illustrated.

Class-Book of Botany—Wood 3 50

Indispensable as a work of reference. Illustrated.

The Laws of Health—Jarvis 1 65

This is not an abstract *anatomy*, but all its teachings are directed to the best methods of preserving health, as inculcated by an intelligent knowledge of the structure and needs of the human body. Illustrated.

Vegetable & Animal Physiology—Hamilton 1 25

An exhaustive analysis of the conditions of life in all animate nature. Illustrated.

Elements of Zoology—Chambers 1 50

A complete view of the animal kingdom as a portion of external nature. Illustrated.

Astronography—Willard 1 00

The elements of astronomy in a compact and readable form. Illustrated.

Elements of Geology—Page 1 25

The subject presented in its two aspects of interesting and important. Illustrated.

Lectures on Natural History—Chadbourne 75

The subject is here considered in its relations to intellect, taste, health, and religion.

VALUABLE LIBRARY BOOKS.

The Political Manual—Mansfield $1 25

Every American youth should be familiar with the principles of the government under which he lives, especially as the policy of this country will one day call upon him to participate in it, at least to the extent of his ballot.

American Institutions—De Tocqueville . . 1 50

Democracy in America—De Tocqueville . . 2 25

The views of this distinguished foreigner on the genius of our political institutions are of unquestionable value, as proceeding from a standpoint whence we seldom have an opportunity to hear.

Constitutions of the United States . . . 2 25

Contains the Constitution of the General Government, and of the several State Governments, the Declaration of Independence, and other important documents relating to American history. Indispensable as a work of reference.

Public Economy of the United States . . . 2 25

A full discussion of the relations of the United States with other nations, especially the feasibility of a free-trade policy.

Grecian and Roman Mythology—Dwight . 2 25

The presentation, in a systematic form, of the Fables of Antiquity, affords most entertaining reading, and is valuable to all as an index to the mythological allusions so frequent in literature, as well as to students of the classics who would peruse intelligently the classical authors. Illustrated.

Modern Philology—Dwight 1 75

The science of language is here placed, in the limits of a moderate volume, within the reach of all.

General View of the Fine Arts—Huntington 1 75

The preparation of this work was suggested by the interested inquiries of a group of young people, concerning the productions and styles of the great masters of art, whose names only were familiar. This statement is sufficient index of its character.

Morals for the Young—Willard 75

A series of moral stories, by one of the most experienced of American educators. Illustrated.

Improvement of the Mind—Isaac Watts . . 50

A classical standard. No young person should grow up without having perused it.

A. S. Barnes & Company

[From the New York Pathfinder, Aug. 1866.]

This well-known and long-established Book and Stationery House has recently removed from the premises with which it has been identified for over twenty years, to the fine buildings, Nos. 111 and 113 William Street, corner of John Street, New York, one block only from the old store. Here they have been enabled to organize their extensive business in all its departments more thoroughly than ever before, and enjoy facilities possessed by no other house in New York, for handling in large quantities and at satisfactory prices every thing in their line.

A visit to this large establishment will well repay the curious. On entering, we find the first floor occupied mainly by offices appertaining to the different departments of the business. The first encountered is the "Salesman's Office," where attentive young men are always in waiting to supply the wants of customers. Further on we come to the Entry Department, where all invoices from the several sales-rooms are collected and recorded. Next comes the General Office of the firm. Then a modest sign indicates the entrance to the "Teachers' Reading-Room"—a spacious and inviting apartment set apart for the use of the many professional friends and visitors of this house. On the table we noticed files of educational journals and other periodical matter—while a book-case contains a fine selection of popular publications as samples. The private office of the senior partner, and the Book-keeper's and Mailing Clerk's respective apartments, are next in order, and complete the list of offices on this floor. The remainder of the space is occupied by the departments of stock known as "Late Publications" and "General School Books."

Descending to the finely lighted and ventilated basement, we find the "Exchange Trade," "Shipping," and "Packing" departments. Here, also, is kept a heavy stock of the publications of the house, while a series of vaults under the sidewalk afford accommodation for a variety of heavy goods. Stepping on the platform of the fine Otis' Steam-elevator, which runs from bottom to top of the building, the visitor ascends to the

Second Story.—This floor is occupied by the Blank Book and Stationery Department, where are carried on all the details of an entirely separate business, by clerks especially trained in this line. Here every thing in the way of imported and domestic stationery is kept in vast assortment and to suit the wants of every class of trade. The system of organization mentioned above enables this house to compete successfully with those who make this branch a specialty, while the convenience to booksellers of making all their purchases at one place is indisputable.

On the third floor are found the following varieties of stock: Toy and Juvenile Books, Bibles and Prayer Books, Standard Works, Photograph Albums, &c. The fourth and fifth stories are occupied as store-rooms for Standard School Stock. During the summer, while all the manufacturing energies of the concern are devoted to the preparation and accumulation of stock for the fall trade, upwards of *half a million of volumes* are gathered in these capacious rooms at once.

The manufacturing department of this house is carried on in the old premises, Nos. 51, 53, and 55 John Street, and 2, 4, and 6 Dutch Street. A large number of operatives, with adequate presses and machinery, are constantly employed in turning out the popular publications of the firm

The Peabody Correspondence.

NEW YORK, *April 29*, 1867.

TO THE BOARD OF TRUSTEES OF THE PEABODY EDUCATIONAL FUND:

GENTLEMEN—Having been for many years intimately connected with the educational interests of the South, we are desirous of expressing our appreciation of the noble charity which you represent. The Peabody Fund, to encourage and aid common schools in these war-desolated States, can not fail of accomplishing a great and good work, the beneficent results of which, as they will be exhibited in the future, not only of the stricken population of the South, but of the nation at large, seem almost incalculable.

It is probable that the use of meritorious text-books will prove a most effective agency toward the thorough accomplishment of Mr. Peabody's benevolent design. As we publish many which are considered such, we have selected from our list some of the most valuable, and ask the privilege of placing them in your hands for gratuitous distribution in connection with the fund of which you have charge, among the teachers and in the schools of the destitute South.

Observing that the training of teachers (through the agency of Normal Schools and otherwise) is to be a prominent feature of your undertaking, we offer you for this purpose 5,000 volumes of the "Teachers' Library,"—a series of professional works designed for the efficient self-education of those who are in their turn to teach others—as follows:—

500 **Page's Theory** and Practice of Teaching.	250 Bates' Method of Teachers' Institutes.
500 Welch's Manual of Object-Lessons.	250 De Tocqueville's American Institut'ns.
500 Davies' Outlines of Mathematical Science.	250 Dwight's Higher Christian Education.
	250 History of Education.
250 Holbrook's **Normal Methods of** Teaching.	250 Mansfield on American Education.
	250 Mayhew on Universal Education.
250 Wells on Graded Schools.	250 Northend's Teachers' Assistant.
250 Jewell on School Government.	250 Northend's Teacher and Parent.
	250 Root on School Amusements.
250 Fowle's Teachers' Institute.	250 Stone's Teachers' Examiner.

In addition to these we also ask that you will accept 25,000 volumes of school-books for intermediate classes, embracing—

5,000 The National Second Reader.	5,000 Beers' Penmanship.
5,000 Davies' Written Arithmetic.	500 First Book of Science.
5,000 Monteith's Second Book in Geography.	500 Jarvis' Physiology and Health.
	500 Peck's Ganot's Natural Philosophy.
3,000 Monteith's United States History.	500 Smith & Martin's Book-keeping.

Should your Board consent to undertake the distribution of these volumes, we shall hold ourselves in readiness to pack and ship the same in such quantities and to such points as you may designate.

We further propose that, should you find it advisable to use a greater quantity of our publications in the prosecution of your plans, we will donate, for the benefit of this cause, *twenty-five per cent.* of the usual wholesale price of the books needed.

Hoping that our request will meet with your approval, and that we may have the pleasure of contributing in this way to wants with which we deeply sympathize, we are, gentlemen, very respectfully yours,   A. S. BARNES & CO.

BOSTON, *May 7*, 1867

MESSRS. A. S. BARNES & CO., PUBLISHERS, NEW YORK:

GENTLEMEN—Your communication of the 29th ult., addressed to the Trustees of the Peabody Education Fund, has been handed to me by our general agent, the Rev. Dr. Sears. I shall take the greatest pleasure in laying it before the board at their earliest meeting. I am unwilling, however, to postpone its acknowledgment so long, and hasten to assure you of the high value which I place upon your gift. Five thousand volumes of your "Teachers' Library," and twenty-five thousand volumes of "School-Books for intermediate classes," make up a most magnificent contribution to the cause of southern education in which we are engaged. Dr. Sears is well acquainted with the books you have so generously offered us, and unites with me in the highest appreciation of the gift. You will be glad to know, too, that your letter reached us in season to be communicated to Mr. Peabody, before he embarked for England on the 1st instant, and that he expressed the greatest gratification and gratitude on hearing what you had offered.

Believe me, gentlemen, with the highest respect and regard, your obliged and obedient servant,   ROBT. C. WINTHROP, Chairman.